D0712142

GEOLOGY EXPLAINED IN
SOUTH WALES

GEOLOGY EXPLAINED IN SOUTH WALES

by
T. R. OWEN

DAVID & CHARLES : NEWTON ABBOT

0 7153 5860 X

Set in 10/12point Pilgrim
and printed in Great Britain
by W. J. Holman Limited Dawlish
for David & Charles (Holdings) Limited
South Devon House Newton Abbot Devon

Dedicated to
my grandchildren
Gareth and Meriel

Contents

Introduction

The aims of this book are threefold. First, it aims to awaken an interest in a subject which can be as much a layman's hobby as a geologist's profession. A broad awareness of the geology of an area can add to one's enjoyment of the countryside. It can make a visit to a seaside resort very much more enjoyable, because one begins to examine the rock layering of the cliffs and to look for fossil bands in the strata. When one finds a fossil in a rock, one may be seeing the remains of an animal that has been hidden from view for hundreds of millions of years. In every adult there is still a boyish thrill in discovering something— an attractive mineral specimen or a good fossil specimen.

Secondly, the book aims to give the reader a broad account of the geology of South Wales, an area noted for its variety of rock types and for the long span of geological time covered by its geological systems—from the Precambrian to the Jurassic. This great geological diversity in an area only one hundred and twenty miles from east to west is the result of the many violent changes in ancient geographies and of the many powerful crustal movements which have occurred. Many of the great geological pioneers of the last century began their studies in Wales or its borderland. Well-known international divisions of geological time are named after Welsh areas. In this account, certain areas have been deliberately highlighted and described in more detail. They are inevitably those areas which have become classic and which have been much visited by students from all over the world. In a well-known fossil locality in North Pembrokeshire or in the Gower Peninsula you may well see a party of geological students from Cologne or from Texas. You will probably also see a party of adults, belonging to an adult education class and interested in the subject only as a pastime.

Thirdly, the book hopes to show that geology is not only a

9

'doing' subject but at the same time a 'thinking' subject. We all inevitably become Sherlock Holmes when we pick up a geological specimen and we ask ourselves the questions 'when', 'how' and even 'why'. Geology is therefore a cultural as well as a practical subject. A study of the past helps us to understand the present, perhaps one day even to forecast the future. Geology tells us about conditions in the past, about life in the past and how life has evolved. It is beginning to tell us about the origins of oceans and how continents have formed and moved. In this account of the rocks of South Wales it is hoped to reveal some of the great changes which have occurred in that area during a very long span of time.

GEOLOGICAL TIME AND THE GEOLOGICAL MAP
OF SOUTH WALES

Geological time is vast, so vast that it is virtually impossible to imagine. It is like trying to think of a distant galaxy as being so many billions of light years away. The Earth probably came into being about 5,000 million years ago and this then is the length of geological time. It is obvious that such a tremendous range of time must be subdivided into many smaller divisions, units that we can handle and to which we can refer. Figure 1 shows the way in which the 'geological column' has been traditionally subdivided. The first major divisions are the four great 'eras': Eozoic, Palaeozoic, Mesozoic and Cenozoic ('dawn', 'ancient', 'middle' and 'recent' life). These divisions are very unequal in length. The Eozoic or earliest era is much longer than the other three added together, in fact it represents almost nine-tenths of geological time. As with historic time, a greater amount of detail enters into the more recent portions and the units of time-range become smaller. Until this century, not too much was known about the Eozoic rocks of the world. Therefore one did not worry too much about subdividing this very long era. In recent years, however, these ancient rocks have been much studied in many countries. Moreover, we now have refined methods of 'dating' rocks, using certain radioactive substances as 'geological clocks'. These materials include uranium, rubidium and potassium. We have, for example, dated rocks in the Outer Hebrides as being 2,600 million years old. The radioactive sub-

FIG 1

(ERA)	(PERIOD OR SYSTEM)	(AGE OF BASE OF PERIOD)	(MAIN EVENTS IN SOUTH WALES)
CENOZOIC	QUATERNARY	2 MILLION YRS	THE ICE AGE
	PLIOCENE	7 " "	FALLING SEA LEVELS
	MIOCENE	26 " "	EARTH MOVEMENTS (ALPINE)
	OLIGOCENE	38 " "	BASIN LAKES AND SWAMPS
	EOCENE	54 " "	UPLIFT AND
	PALAEOCENE	65 " "	EROSION
MESOZOIC	CRETACEOUS	136 " "	"THE CHALK SEA" UPLIFT AND SLIGHT EARTH MOVEMENTS
	JURASSIC	194 " "	JURASSIC SEAS
	TRIASSIC	225 " "	DESERT BASINS AND SEAS WITH UPLANDS
PALAEOZOIC	PERMIAN	280 " "	UPLIFT AND GREAT EROSION
	CARBONIFEROUS	345 " "	ARMORICAN EARTH MOVEMENTS FORESTED SWAMPS CLEAR SEAS
	DEVONIAN	395 " "	ESTUARIES, MUDFLATS AND DELTAS
	SILURIAN	435 " "	CALEDONIAN EARTH MOVEMENTS SEAS IN PLACES
	ORDOVICIAN	500 " "	SEAS AND VOLCANOES
	CAMBRIAN	600 " "	SEAS
EOZOIC	PRECAMBRIAN	ORIGIN OF THE EARTH 5000 MILLION YEARS AGO	MARINE INVASION PROLONGED VOLCANIC ACTIVITY MANY EARTH MOVEMENTS ? ? ?

stances are subject to breakdown and the rate of breakdown is known, and known to be constant for any one substance. This 'radiometric' dating has been carried out on rocks of every geological era and we now know the approximate dates of commencement of the three smaller eras. The Palaeozoic era began 600 million years ago. Before this, in the long Eozoic era, life was probably soft-bodied and fossils were only rarely preserved. At the beginning of the Palaeozoic, animals began to develop hard outer coverings or shells which could be preserved as fossil remains. The Mesozoic era commenced about 225 million years ago. Life in this era was dominated by the coiled ammonites and by the great reptile group known as the dinosaurs. The beginning of the last era, at about 65 million years ago, saw the rapid development and evolution of the mammals, leading ultimately to the emergence of man.

The three post-Eozoic eras are each subdivided into a number of 'periods'. The Palaeozoic is divided into six periods, three in the Lower Palaeozoic and three in the Upper Palaeozoic. Three periods make up the Mesozoic and six occur in the Cenozoic, though these six periods are so short, by comparison with earlier periods, that many geologists combine them.

During each one of the geological periods, several thousands of feet of sediments have been deposited in basins, vast seas, deserts, estuaries, lakes, swamps, deltas, etc. With time, and burial by younger sheets of sediment, the deposits harden (by compaction, crystallisation, drying processes, etc) into rock layers or 'strata'. Thus soft wet muds harden into shales, clays and tough mudstones; sands become sandstones or grits and pebbly deposits become 'conglomerates'. Whereas a 'period' refers to a length of time, a geological 'system' refers to the rocks formed during that time. Thus to talk about 'the Carboniferous Period' is to refer to some 65 million years of time. 'The Carboniferous System', on the other hand, concerns some 15,000ft of limestones, grits, shales, coals and sandstones, deposited during that 65 million years.

Rocks which before compaction and hardening were soft, plastic or friable sediments are known as 'sedimentary rocks'. They include clays, mudstones, shales, sandstones, grits, limestones (of many kinds), breccias (one-time screes) and conglomerates (gravels and pebbly sands). Besides being layered or 'stratified', they are usually fossiliferous to a greater or lesser degree, though some can be barren, especially those deposited during the Eozoic era (Precambrian rocks). Fossils are the remains or traces (footprints, moulds, casts, trails, etc) of once-living animals or plants. They are best preserved in fine-grained sediments. In coarse breccias and conglomerates they have usually been destroyed or broken up. Different fossil assemblages characterise each era and even each period or system. Thus 'graptolites' characterise the Ordovician and Silurian systems, but those of the former are different from those found in the latter system or period. It is possible, on a fossil basis, to even further subdivide the periods or systems into smaller and smaller units. Sometimes even a thin group of beds, no more than a few feet thick, can be identified by its fossils, as for example certain bands in the Millstone Grit or the Coal Measures. The same bands with the same fossils may then even be known in the British Isles and in other European countries and correlations can then be made across large areas. Furthermore, finding those distinctive fossils in the core of a deep borehole will help to identify rock layers perhaps deep down below the land surface or sea bottom. Those fossil bands may lie near to certain porous sandstones known to be good reservoirs for oil or natural gas, or near to a thick group of valuable coals. The geologist is forever aiming at more detailed subdivisions.

It should be pointed out that not all the rocks are of such sedimentary origin. From time to time in the geological past (and even today), hot molten material ('magma') pushes up through the upper regions of the earth's crust. Some of this hot magma reaches the surface and flows out over the land (or sea bed) from a volcano, forming sheets of lava. Sometimes the action is more explosive and ashes or even bomb deposits (agglomerates) accumulate). Not all the magma cools and hardens on the surface. Some of it crystallises as sheets, pipes or larger masses below the surface. The sheets are called 'sills' and the pipes 'dykes' (Figure 2). Larger masses of coarse, crystalline rock are known as bosses, laccoliths and batholiths (these may be many miles in dimension). The deeper the level at which the molten magma cools, the coarser the degree of crystallisation. 'Gabbro' and 'granite' are particularly coarse-grained with individual crystals perhaps up to two inches long. The hardened, crystallised magma forms rocks known as 'igneous'. Surface-cooled lavas and ashes are the 'volcanic' igneous rocks, whilst the deep-seated, cooled products in bosses and batholiths are the 'plutonic'

FIG 2

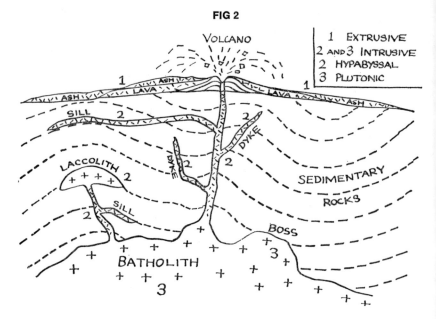

varieties. The intermediate-level dykes and sills are finer in grain than the granites and gabbros of these deep bodies. A dark rock called 'dolerite' is commonly found in dykes and sills. A fine-grained dark lava called 'basalt' is often the result of surface vulcanicity. The Giants Causeway in Northern Ireland is formed of basalt. The lava sheets here have cooled to form six-sided columns.

Two important questions arise at this point. First, if horizontal sheets of sedimentary rocks have in the past been deposited on top of one another, why is it that today we can see the older layers in some areas and younger strata in others? Secondly, if 'granite' and 'gabbro' masses crystallised at depth, why can they be seen on the surface (for example in Cornwall) today? The two questions can be answered together (Figure 3). From time to time in the geological past, great crustal disturbances or earth movements have occurred. Different parts of the world have been affected at different times, but the results have been the folding and 'faulting' (fracturing) of the rock sheets (with their igneous intrusions) formed up to the time of the earth movements. Uplift and intense erosion has usually fol-

FIG 3

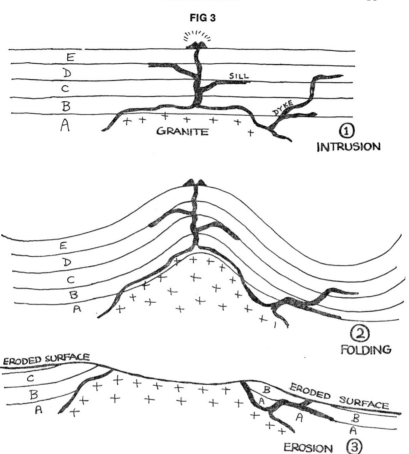

lowed the earth storms and over those regions where the strata have been appreciably upfolded, many thousands of feet of strata have been removed, the waste debris being carried off elsewhere to start new cycles of deposition—to make new sedimentary sheets. This erosion of upwarped areas can therefore result in the exposure, on the erosive surface, of strata of varying ages. Upfolded areas therefore tend to expose the older rocks whereas downwarped belts tend to preserve the younger sheets. Along the axis of a great upfold along the Towy Valley in South Wales, Ordovician rocks are exposed. To the south-east, in the great downfold which is the South Wales coalfield, much

FIG 4

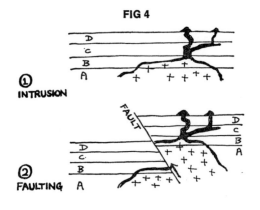

① **INTRUSION**

② **FAULTING**

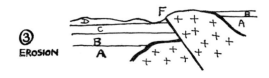

③ **EROSION**

FIG 5

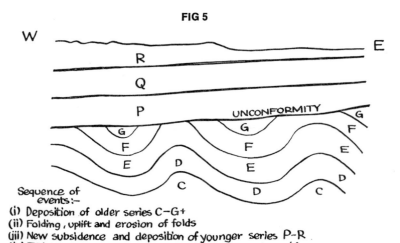

Sequence of events :-
(i) Deposition of older series C–G+
(ii) Folding, uplift and erosion of folds
(iii) New subsidence and deposition of younger series P–R
(iv) Tilting towards the west; erosion of present day surface

younger rocks belonging to the Upper Carboniferous are preserved (Figure 6). It can be seen from Figure 3 that deep-seated plutonic igneous masses, at one time deeply buried, can also be brought upwards to levels of active erosion by the same folding processes. Large-scale fracturing, with powerful upward and downward dislocations, can also help in bringing strata and igneous masses that originally were at fairly deep levels up to the zone of intense surface erosion (Figure 4). Upfolds are called 'anticlines' and downfolds are 'synclines'. Along some faults, however, the movements have been horizontal. Such faults are called 'tear faults' or 'wrench faults'. In Eozoic times there were at least four or five known times of great earth movement. Since then there have been three such great 'orogenies', the Caledonian Orogeny of the Ordovician and Silurian, the Hercynian or Armorican Orogeny (at the close of Carboniferous times), and the Alpine Orogeny in the Miocene period. Because of these various earth movements, great breaks or gaps occur (called 'unconformities') in the total rock successions of any region or country. The rock layers above the unconformity will be much less deformed or folded than those below it (Figure 5).

The pattern of the rock strata, when viewed from above, can therefore be a complex one, especially when the area embraces rocks which have been much folded and faulted, then eroded, and then buried under newer sheets which in their turn have undergone folding and part erosion. In some parts of the area the older strata are exposed on the surface whereas in others they are buried from view by younger layers. This is what a geological map is all about. It shows what rock units occur *on the surface* together with the position of the faults which can dislocate them. The status of the unit varies with the extent of the area depicted on the map. Most regional maps show the various systems and even their immediate subdivisions. In the map of a coalfield area, the surface position or 'outcrop' of a coal seam may be shown. The surface extents of igneous rocks are shown in areas where they occur. Geological maps of British areas may be purchased. Some details of South Wales maps are given on p200.

The general geology of South Wales is shown in Figure 6. It can be seen that the oldest (Eozoic or Precambrian) rocks occur only in Pembrokeshire, as also do the succeeding Cambrian

FIG 6

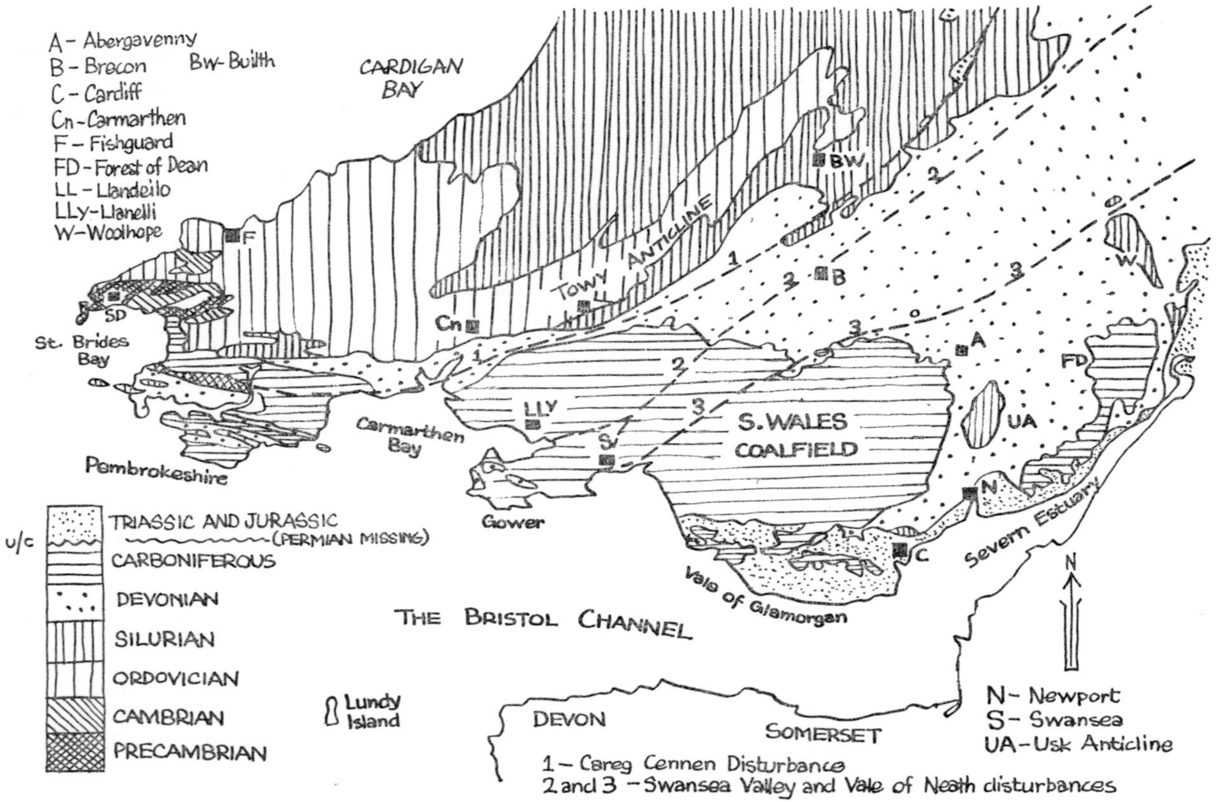

A – Abergavenny   Bw – Builth
B – Brecon
C – Cardiff
Cn – Carmarthen
F – Fishguard
FD – Forest of Dean
LL – Llandeilo
LLy – Llanelli
W – Woolhope

CARDIGAN BAY

TOWY ANTICLINE

St. Brides Bay

Pembrokeshire

Carmarthen Bay

Gower

S. WALES COALFIELD

Vale of Glamorgan

Severn Estuary

THE BRISTOL CHANNEL

SOMERSET

DEVON

§ Lundy Island

∷∷ u/c	TRIASSIC AND JURASSIC
═══	CARBONIFEROUS (PERMIAN MISSING)
∴∴	DEVONIAN
‖‖	SILURIAN
┼┼	ORDOVICIAN
╱╱	CAMBRIAN
▨	PRECAMBRIAN

N – Newport
S – Swansea
UA – Usk Anticline

1 – Careg Cennen Disturbance
2 and 3 – Swansea Valley and Vale of Neath disturbances

rocks. Ordovician rocks occur in Pembrokeshire, Carmarthenshire and Cardiganshire and also north-eastwards into Breconshire, in mid-Wales. Ordovician rocks are often associated with the overlying Silurian in those counties. Devonian (Old Red Sandstone) strata have their greatest surface extent in Breconshire and Monmouthshire but they are also found along the rim of the South Wales Coalfield in Carmarthenshire and Glamorgan and as scattered outcrops in South Pembrokeshire. The main Carboniferous areas are of course the three South Wales coalfields, the central, main coalfield being much larger than the Pembrokeshire and Forest of Dean fields. Other (lower) Carboniferous rocks outcrop extensively in South Pembrokeshire, Gower and in the Vale of Glamorgan.

Permian rocks are unknown in the principality, but Triassic strata are found in the Vale of Glamorgan with very small remnants elsewhere. The still younger Jurassic rocks are restricted (on land) to the Vale of Glamorgan, though we now know that thicker Jurassic successions lie beneath the waters of the Bristol Channel and Cardigan Bay. Apart from the very small patch of Oligocene(?) at Flimston in South Pembrokeshire, no Tertiary deposits are to be found in South Wales. Extensive covers of Quaternary (mainly glacial) deposits occur, however, though the largely unconsolidated state of these youngest sediments hardly warrants them being called rocks. One significantly absent system in Wales is the Cretaceous, in marked contrast to southern and eastern England where rocks like the Chalk and the Greensand are a marked feature of the surface geology. One of the important questions in connection with the geological history of Wales is: did a cover of Chalk once exist? There is now reason to believe that it once did, though probably not for long.

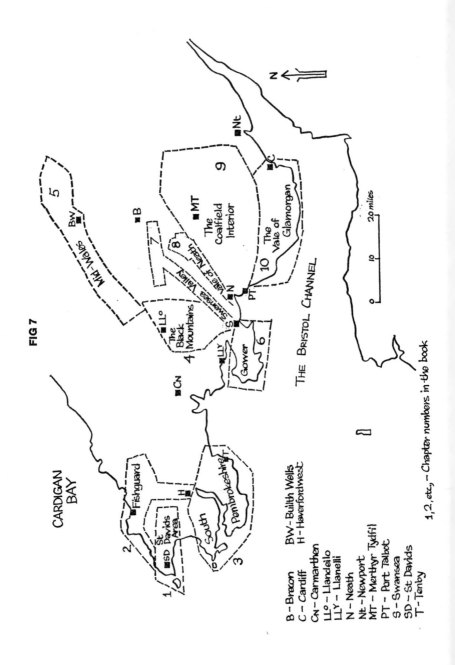

FIG 7

CARDIGAN BAY

THE BRISTOL CHANNEL

Mid-Wales

The Black Mountains

Swansea Valley

Vale of Neath

The Coalfield Interior

The Vale of Glamorgan

Gower

Pembrokeshire

St Davids Area

South

Fishguard

0   10   20 miles

B – Brecon
C – Cardiff
Cn – Carmarthen
LLo – Llandeilo
LLY – Llanelli
N – Neath
Nt – Newport
MT – Merthyr Tydfil
PT – Port Talbot
S – Swansea
SD – St Davids
T – Tenby

BW – Builth Wells
H – Haverfordwest

1, 2, etc., – Chapter numbers in the book

CHAPTER 1

The Land of St David

Pembrokeshire is a geologists' paradise. Its cliff sections reveal rocks covering virtually half of the geological column, from the Precambrian to the Carboniferous, all in an area thirty miles square. The cliff sections in this county are magnificent and the area is made even more attractive to the visitor by having wonderful sandy bays, imposing headlands, deeply penetrating inlets and delightful seaside villages and larger resorts, such as Tenby and Saundersfoot. There is much to interest the historian and the archaeologist. The presence of St David's Cathedral makes this the capital county even if not the capital town of Wales. Castles abound, including the fine examples at Pembroke, Carew and Manorbier. The modern monastery on the Island of Caldy, near Tenby, reminds one that this island had an ancient monastic cell as far back as the sixth century. The map of Pembrokeshire is covered with the remains of ancient Iron Age forts, camps and burial chambers. There are no less than three 'Castell Cochs' (red castles) along the eight miles of coast north-east of St David's. Last but not least, the Prescelly Hills in the north-east of the county were the original home of the famous 'bluestones' of Stonehenge. How these stones got to Stonehenge is a matter for argument and will be discussed later.

From the geological standpoint (Figure 8) Pembrokeshire can be divided into two halves, the dividing line running east-west through the county town of Haverfordwest. The northern area is noted for its Precambrian, Cambrian and Ordovician rock successions. Moreover, both sedimentary and igneous rocks occur, vulcanicity having been widespread at times. The southern area is known best for its Old Red Sandstone and Carboniferous successions and for the spectacular folding and faulting of these strata. Precambrian, Ordovician and Silurian rocks do occur even

21

FIG 8

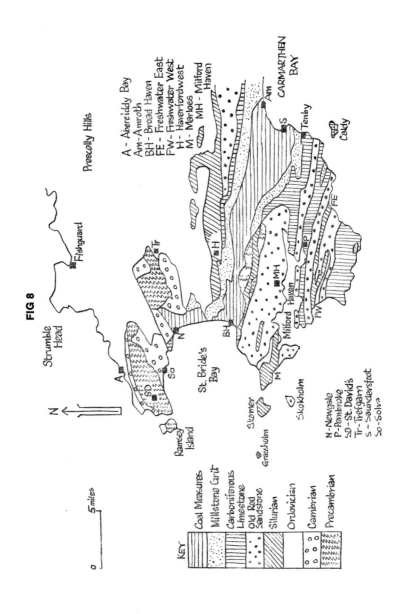

A - Abereiddy Bay
Am - Amroth
BH - Broad Haven
FE - Freshwater East
FW - Freshwater West
H - Haverfordwest
M - Marloes
MH - Milford Haven

N - Newgale
P - Pembroke
SD - St. Davids
Tr - Trefgarn
S - Saundersfoot
So - Solva

Preselly Hills

Strumble Head

Fishguard

CARMARTHEN BAY

Tenby

Caldy

St. Bride's Bay

Ramsey Island

Skomer

Grassholm

Skokholm

Milford Haven

KEY

	Coal Measures
	Millstone Grit
	Carboniferous Limestone
	Old Red Sandstone
	Silurian
	Ordovician
	Cambrian
	Precambrian

0   5 miles

in this southern area. This is because of the intensity of the
buckling and fracturing. The Islands of Skomer and Grassholm
(famous bird sanctuaries) are formed of thick Silurian volcanics.

The coastal outline of Pembrokeshire is dominated by three
features, the great projecting peninsulas of St David's and
Marloes-Skomer and the deep drowned inlet of Milford Haven.
The two peninsulas enclose the great square-shaped St Brides
Bay, the northern coast of which contains the lovely ria of Solva
and numerous bays south of St David's. The northern coasts of
the St David's Peninsula are jagged and high with hard igneous
rocks forming the headlands and the less resistant slate formations being eroded to form intervening bays, as at Abereiddy.
The great Milford Haven, fed by the two Cleddau rivers, is a
drowned valley system submerged under the rising waters of
Neolithic to Bronze Age times. At the entrance to the Haven the
Dale Peninsula projects southwards, its red cliffs reflecting in
Pembrokeshire's blue-green waters. The south coast of the Haven
is indented by the almost circular Angle Bay, during the war a
haven for flying boats, and by the estuary of the Pembroke
River.

The county is largely relatively low-lying with large areas,
especially south of the Haven, at 200 feet (or lower) above sea
level. There appear to be three main height levels at 200, 400 and
600 feet above sea level. The remnants of the 600 feet surface
are now either isolated hills, like Carn Llidi, or elongated narrow
ridges, such as the ridge of tough Precambrian volcanics extending from Roch to Trefgarn (or Treffgarne). North-eastwards,
the land rises appreciably into the Prescelly Hills, an area of
Ordovician sediments and volcanics, invaded by many intrusions
of resistant dolerite (the source of the Stonehenge blue stone).
Heights of over 1,000 feet are reached, as for example on
Mynydd Castlebythe (1,137 feet) and Cerriglladron Foeleryr
(1,535 feet). It is in these hills that the feeders of the southward-
flowing Cleddaus, and the northward-flowing Gwaun, rise.

THE ST DAVID'S PENINSULA

This major Pembrokeshire headland is, from a geological
standpoint, dominated by the St David's Anticline, an almost
E-W upfold which brings Precambrian rocks to the surface,

FIG 9

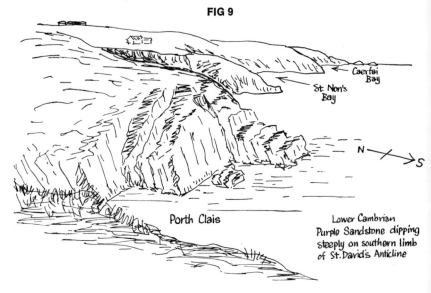

Porth Clais

Caerfai Bay

St Non's Bay

N / S

Lower Cambrian Purple Sandstone dipping steeply on southern limb of St. David's Anticline

especially around the cathedral city. On the southern flank of the anticline, Cambrian rocks dip steeply southwards along the south coast of the peninsula. These rocks are magnificently displayed in the bays and headlands of that coastal fringe (Figure 9). On the north coast of the peninsula, Ordovician strata, much intruded by igneous material, dip northwards. There are many faults, trending mainly N-S and E-W. The latter directed fractures are often lengthy, with big throws. One well-marked example trends across the bays of Caerbwdy, Caerfai and St Non's. In Caerbwdy Bay the fault is associated with an overturned fold, well seen in the eastern cliffs of that bay when viewed from the western side. The city itself is situated on Precambrian intrusives known as the Dimetian, intruded into a thick series of Precambrian volcanics (ashes, lavas and tuffs) known as the Pebidian. The cathedral is built on a Dimetian quartz-porphyry, extending in a NNE-SSW direction but cut off on the southern side by one of the major E-W faults. A 'porphyry' is an igneous rock in which there were two stages of cooling of the molten material in rocks not quite at the surface. In the initial slow cooling, large crystals were able to form (of quartz, in this St David's example) but then the rest of the molten 'soup' cooled more rapidly, forming a 'groundmass' of fine-grained intergrowths of quartz, felspar and mica. If all the cooling had proceeded slowly, the resulting rock would have been a granite.

The cathedral, however, is largely built of purple Lower Cambrian sandstone, named after Caerbwdy Bay. The cathedral is probably the oldest in the British Isles, with the longest line of bishops—one hundred and twenty-three. The traces of the first tiny abbey church of Dewi (St David) around AD 520 have all vanished. William the Conqueror visited St David's, where the present building was begun in 1180. The nave is thirteenth century, built by half-Norman, half-Welsh bishops. The main and finest builder was Bishop Henry Gower. Much of the building was rebuilt and refurnished in the early sixteenth century—the marvellous Irish bog oak roof in 1500, the bell tower at about the same time. Subsidence caused part of the nave and western end to fall in the eighteenth century and rebuilding at about 1789 was by the great John Nash. More of the purple stone was used at this time. The nearby Bishop's Palace, with its use of multicoloured stone, was chiefly the work of Henry Gower. Thieves robbed this building of its lead roof.

St Non's Bay. This bay is worth visiting in order to examine the Pebidian volcanic rocks, well exposed in its innermost cliffs, and also to see the junction of those Precambrian rocks with the succeeding Cambrian basal conglomerate. The bay is reached by taking the narrow road out of the west side of the city towards the Warpool Court Hotel and then turning towards a small monastery. Nearby is the older ruined monastery and also a wishing well, believed to have been formed during a lightning flash at the time of the birth of St David. From here a path descends the cliffs via a deep cleft (marking a fault). In the walls of this gash are exposed the Precambrian volcanics. Here these are banded rhyolite lavas (light-coloured silica-rich rocks) and silicified ashes known as 'halleflinta'. The west side of the cleft also shows a sheared dolerite intrusion. A little way seawards on the rocky foreshore, an elevated feature forming a line of islands at high tide exposes the pebbly base of the Cambrian System. This reddish coloured 'conglomerate' is massive but the beds are tilted slightly beyond the vertical. Note that the underlying Pebidian volcanics are also stained red or purple. This could represent the original weathered colour of the old eroded Pre-

FIG 10

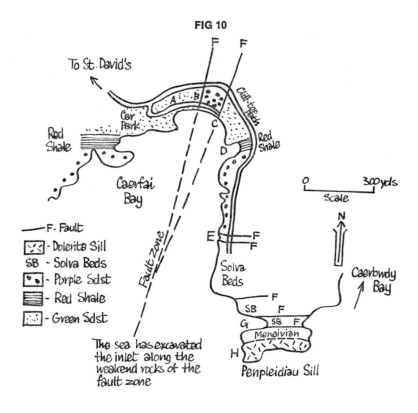

To St. David's

F F

A B

Car Park

Cliff-beach

Red Shale

C

Red Shale

D

Caerfai Bay

0 ——— 300 yds
Scale

N

Caerbwdy Bay

— F - Fault
⟨∨∨⟩ - Dolerite Sill
SB - Solva Beds
⟨•∘•⟩ - Purple Sdst
▦ - Red Shale
⟨∷∷⟩ - Green Sdst

Fault zone

The sea has excavated the inlet along the weakened rocks of the fault zone

E F
F

Solva Beds

F

SB F

G SB F

Menevian

H

Penpleidiau Sill

FIG 11

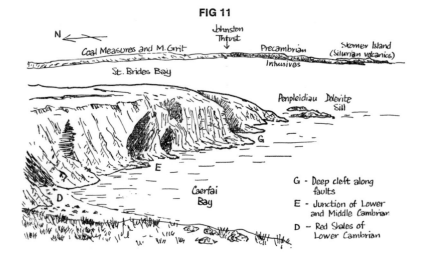

N

Johnston Thrust

Coal Measures and M. Grit

Precambrian Intrusives

Skomer Island (Silurian volcanics)

St. Brides Bay

Penpleidiau Dolerite Sill

G

E

Caerfai Bay

D

G - Deep cleft along faults
E - Junction of Lower and Middle Cambrian
D - Red Shales of Lower Cambrian

cambrian land surface, prior to the great marine invasion by the Cambrian sea. In a cove immediately to the east another intrusion of quartz-porphyry can be seen.

Caerfai Bay In this inlet, one mile to the east, the remainder of the Lower Cambrian and some of the Middle Cambrian succession can be seen. The bay is approached from the eastern end of St David's by taking the narrow road past the Twr y felin Hotel. Cars may be parked near the top of the path which descends into the bay. Figure 10 shows the geology of part of the bay and points of interest are marked on this map and on the view shown in Figure 11. The bay is there because the sea has eroded along a fault zone where the rocks have been crushed and weakened by the movements along the fractures. Water is seen to be seeping along both faults and this, of course, aids the erosive processes.

With the exception of the basal conglomerate seen at St Non's, the whole of the Lower Cambrian succession is seen on the eastern side of Caerfai Bay. Above the conglomerate (indifferently exposed north of the bay) comes a sequence of green sandstone beds, weathering rather pale green or yellow. They are seen on the first part of the descent path (**A** in Figure 10). The highest beds of this green formation become purple to red in colour (as at **D**) and this marks the passage into a 40 foot sequence of rather bright red shales. The softer character of these shales accounts for the recesses worn into them on both the eastern and western cliffs of the bay. The shales show ripple-marking and worm burrowing. Thin bands of volcanic ash can also be seen. Above the Red Shales come the purple Caerbwdy sandstones, some 500 feet thick.

The junction of the Lower and Middle Cambrian occurs in the cave complex at **E** in Figures 10 and 11. Some people claim the (Middle Cambrian) Solva Beds rest with discordance on the Caerbwdy Sandstone, but faulting at this point makes it difficult to decide. The Solva Beds (green pebbly sandstones with green and purple mudstones and shales) reach virtually to the headland, but in a deep, fault-bound cleft (at **G**) much higher Middle Cambrian (Menevian) mudstones come on, invaded by an igneous sill formed of dark dolerite and known as Penpleidiau. From this headland it is well worth walking a little way round into Caerbwdy Bay in order to glimpse the remarkable overfold

and fault on the eastern cliffs. This disturbance must be continuous with the fault complex at **E** in Caerfai Bay. From the
headland too there are superb views across St Brides Bay and
round to Skomer. On days of clear visibility the island of Grassholm is seen on the horizon. The eastern cliffs of St Brides Bay
are carved into Coal Measures and Millstone Grit, and the form
of the great bay suggests that the same soft measures must form
large parts of its floor. The sea has eroded deeply into these
softer beds, leaving two hard prongs to the north and south:
the northern prong is made up of the Precambrian and Cambrian of St David's whilst the southern arm is formed of tough
Precambrian intrusives and (further west) the Silurian volcanics
of Skomer.

One further point about St Brides Bay should be pointed out.
A veneer of glacial boulder clays occurs in the bay, and the
glacial pebbles are being continually winnowed out by the sea
and are being driven by the south-west winds and tides into the
north-easternmost corner of St Brides Bay, that is towards the
village of Newgale. The result is the superb 'storm beach' of
boulders and shingle forming a seaward rampart to that village.
The pebble ridge holds back the stream, forming a marsh inland.
The ridge is continually moving inland, and is close to the road.
It is said that the local inn has had to 'move inland' on more
than one occasion.

Solva   Midway between Caerfai Bay and Newgale is the magnificent inlet of Solva, where the River Solfach reaches the sea.
In its higher course this river flows westwards, and it is thought
that at one time the river flowed into Whitesands Bay, as did
also the River Alun of St David's. Now both rivers turn sharply
southwards, probably having been forced to do so because of
glacial barriers or even by a wall of retreating Irish Sea ice near
the close of the Ice Age (some eleven thousand years ago).

The Solva inlet is a drowned valley or 'ria'. After the Ice Age
the River Solfach probably continued in its valley well out into
St Brides Bay. As the sea level rose, following on the return
of melted ice-water to the sea, the lower reaches of the valley
became drowned. This drowning (see Chapter 10) lasted from
Neolithic to Bronze Age times and a total drowning of about
80 feet occurred. The legends of drowned settlements along the
coasts of West and North Wales are not altogether without
some geological foundation.

FIG 12

W

E

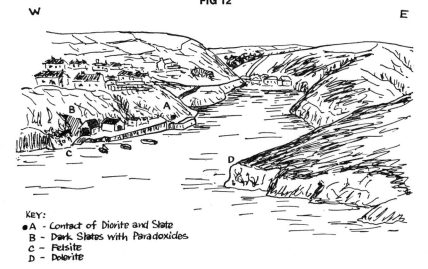

KEY:
- A - *Contact of Diorite and Slate*
- B - *Dark Slates with Paradoxides*
- C - *Felsite*
- D - *Dolerite*

The Solva ria has an inner NE-SW portion, defined by a fault zone of the same trend, and an outer N-S entrance. The steep seaward-dipping Cambrian sandstones, flags and cleaved mudstones are invaded by at least three kinds of intrusions—quartz diorite, felsite and dolerite. The diorite, fairly coarse grained with crystals of dark hornblende and whitish felspar, can be easily traced on the right bank of the inlet midway between the car park and the lifeboat station (**A** in Figure 12). The two edges of this intrusion can be traced, showing the baking ('metamorphism') of the Cambrian strata at the intrusive margins. At the lifeboat station there is a contact of dark blue (Menevian) cleaved mudstones with a porphyritic felsite intrusion. The mudstones have yielded trilobites such as the large *Paradoxides davidis* (Figure 15). (Particularly large specimens of this the largest trilobite have been found at Porth y Rhaw, one and a half miles west of the Solva ria. A large specimen can be seen in the National Museum of Wales, Cardiff.) The outer cliffs of the Solva inlet are carved into the (Upper Cambrian) Lingula Flags (described below), but yet another intrusion, this time of dolerite, marks the actual mouth of the ria.

Whitesand Bay This large bay on the west coast of the St David's Peninsula is about three miles by road from the city and there is a large car park. Dominating the sandy bay is the hill of Carn Llidi (595 feet OD), one of several intrusions of a coarse grained 'gabbro' in this area—St David's Head is another. Whitesand Bay is carved into Cambrian strata and lower Ordovician slates; Figure 13 shows the geology of the bay. Prominent headlands occur at the points marked **A** to **D**. **A** and **B** are caused by

FIG 13

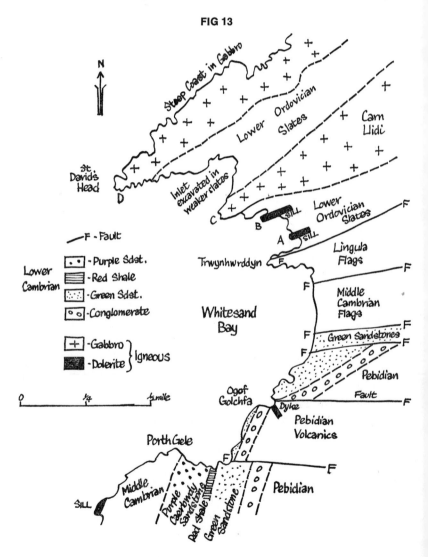

the resistant sills of dolerite, standing vertical in vertical Ordovician slate. **C** and **D** are headlands of coarse, extremely hard gabbro. The most remarkable headland however is that called 'Trwynhwrddyn' (Welsh for 'The Ram's Nose'). This is a headland formed of alternations of thin sandstones with shales and normally would not be expected to make a resistant headland. In this exceptional case, however, a fault zone has been riddled with quartz veins (probably carried upwards by hot igneous vapours) and these veins have strengthened the resistance of the strata. A panoramic view of the headland (and of the dolerite **A**) *is shown in Figure* 14.

The basal conglomerate of the Cambrian is excellently exposed in a minor inlet called Ogof Golchfa ('washing place cave') on the southern side of Whitesand Bay. The conglomerate once again overlies Pebidian tuffs which are reddened. A few feet to the east, the volcanics are intruded by a massive, dark greenish dolerite. There are many E-W faults in Whitesand Bay and one of them can be plainly seen in the innermost recess of Ogof

FIG 14

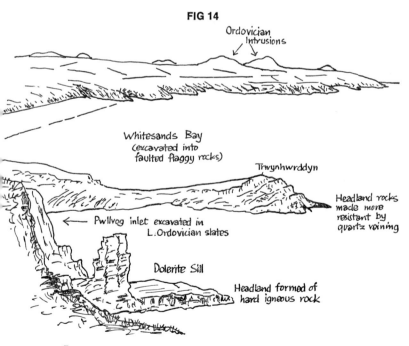

Ordovician Intrusions

Whitesands Bay (excavated into faulted flaggy rocks)

Trwynhwrddyn

Headland rocks made more resistant by quartz veining

← Pwllveg inlet excavated in L.Ordovician slates

Dolerite Sill

Headland formed of hard igneous rock

B

Golchfa. There are vertical 'slickensides' (polished grooves) on the fault face. They show the direction of slip along the fault. The exact position of the strata of the middle cliffs of Whitesand Bay within the Cambrian is not known, because of their un-fossiliferous character. The strata immediately south of the Trwynhwrddyn headland are however quite definitely the (Upper Cambrian) 'Lingula Flags'. These are rapid alternations of sandstones, siltstones and shales. The sandstones show 'graded-bedding' in which the sand grains become finer grained upwards within any one bed. This is one way in which one can tell whether a bed is the right way up or whether it is inverted. On some of the bedding planes of the more shaley strata, small black-shelled brachiopods called *Lingulella* can be found. This is one of the first brachiopods to appear in the fossil record. *Lingula* still lives today and has changed very little through over 500 million years of time. Beyond Trwynhwrddyn, the bay (Pwlluog) between that headland and the dolerite sill (**A**) exposes very steeply dipping dark slates, weathering rusty. These are the lowest beds of the Ordovician here and have yielded occasional specimens of graptolites such as *Tetragraptus* (Figure 15). The cliffs above the eastern end of Trwynhwrddyn show glacial boulder clay. A search of the various boulders is rewarding, because besides local rocks such as the Ordovician gabbro and dolerite one can sometimes find rocks from North Wales or even from areas bordering the north Irish Sea, for example granite from Northern Ireland. These boulders and pebbles were brought down the Irish Sea depression by a great Irish Sea ice sheet. On one occasion, Irish Sea ice moved up the Bristol Channel, almost to Cardiff.

The Carn Llidi intrusion extends seawards to the headland **C**. The gabbro is well exposed on the top surface of the headland. The margins of the intrusion can be seen to be finer grained than the central portion. This is because the margin cooled more quickly because of the chilling contact of the molten material with the rocks into which it was intruded. The gabbro which forms St David's Head is very similar in character to the Carn Llidi intrusion. It could be that faulting has in fact repeated the gabbro outcrop.

Abereiddy Bay This bay is reached by taking a minor road from St David's to the village of Llanrian but turning left at a

FIG 15

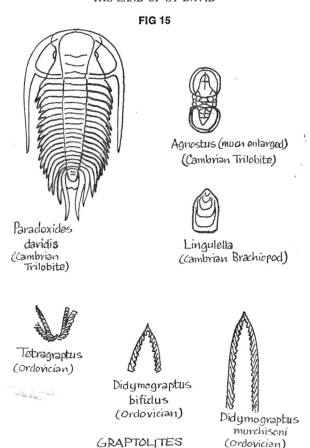

Paradoxides
davidis
(Cambrian
Trilobite)

Agnostus (much enlarged)
(Cambrian Trilobite)

Lingulella
(Cambrian Brachiopod)

Tetragraptus
(Ordovician)

Didymograptus
bifidus
(Ordovician)

Didymograptus
murchisoni
(Ordovician)

GRAPTOLITES

crossroads one mile south-west of the village. It can also be reached from the main A487 road at Groes goch. In Abereiddy Bay are exposed rocks belonging to a division of the Ordovician System known as the Llanvirn Series, after a farm called Llanvirn, just south of the bay. In Pembrokeshire, the Llanvirn Series consists mainly of dark slates and because here the bedding of the strata coincides with the cleavage or splitting imposed on them by earth movements, the slates are very fossiliferous, graptolites being particularly abundant. The Llanvirn graptolites are of the 'tuning fork' variety, with two branches (or 'stipes') hanging downwards (Figure 15). In the lower Llanvirn slates the form *Didymograptus bifidus* is found but in the higher beds the larger *Didymograptus murchisoni* occurs abundantly. A very good locality for finding it is at the road junction above the innermost south-east corner of the bay (after turning a sharp bend). The straight form *Glyptograptus* also occurs. A little way down the road towards the sea, the slates can be seen to be very contorted in the road banks. This is due to frost shattering and to 'creep' down slopes at a late stage in the Ice Age. At that time, alternate thawing and freezing was particularly active, and when surface thawing occurred, sludging or slipping down slopes occurred on top of still frozen subsurface layers. The process is known as 'solifluction'.

The *bifidus* and *murchisoni* slates of the Llanvirn Series in the Abereiddy region are separated by a volcanic group of rhyolite lavas and ashes (the Llanrian Volcanic Series). These harder rocks form the projecting headland called Trwyncastell ('castle nose'). A good face of these lenticular volcanic sheets, interbedded with slates, can be seen in the small harbour inlet eroded into the headland. A path which skirts round to reach this harbour has one very narrow crossing place. This is the position of a major fault which repeats the succession of the south arm of Abereiddy Bay on this Trwyncastell headland. The slate tips in the inner-most recess of the harbour (used for transporting roofing slates prior to World War I) are worth examining. Besides *bifidus* graptolites, one can find gastropods, a brachiopod called *Monobilina*, nautiloid cephalopods and even trilobites.

Chapter 2

From Ancient Volcanoes to Stonehenge

The great inlet of Milford Haven is fed by the Cleddau rivers, on the westernmost of which stands the county town of Haverfordwest. In earlier times this was the lowest fording point across the river and its castle was built by the Norman conquerors, more particularly Gilbert de Clare, Earl of Pembroke. It was badly damaged during the Civil War. Much of what is now seen was the county gaol but is now the county museum.

Geologically, Haverfordwest is noted for its sections in the Lower Silurian, the best known occurring near the town's gasworks. In fact, the formations here are known as the Gasworks Sandstone and Gasworks Mudstone. The mudstone is seen alongside the gasometer and is here very fossiliferous. Some concern has been felt recently about the spreading of rock debris across the road by fossil hunters and one should therefore take care here. There is no need to hammer the rock beds as there are abundant fossils in the scree. The brown mudstone yields an abundance of brachiopods such as *Dolerorthis, Leptaena, Leptostrophia* and *Plectatrypa*. Trilobites include *Encrinurus*. Corals include *Favosites* and *Petraia*. The long screw-like *Tentaculites* (a gastropod?) is very abundant.

The narrow belt of Silurian rocks at Haverfordwest gives way northwards to a wide belt of Ordovician strata, stretching all the way to Fishguard. This Ordovician tract is, however, punctured by a complex ENE-WSW belt of still older rocks (Cambrian and Precambrian) around the small village of Treffgarne (from 'Trefgarn'—the 'fortress village'). Moreover, the Ordovician slates are, in this neighbourhood, interbedded with the thick Treffgarne Volcanic Series. These essentially explosive volcanics, of andesite composition, give rise to craggy higher ground, well wooded on the eastern side of the river. Brunel's railway line to

Fishguard runs through this area and this famous engineer's name is given to a rock formation (Brunel Beds) at the base of the Ordovician of this region.

The Treffgarne Volcanic Series is excellently exposed in a series of quarries on the west side of the main A40 road (northeast of the village). The southernmost quarry shows the purple and green ashes together with some interbedded shales. Of greater interest, however, are the large bombs (some several feet across) seen in the northern face of this quarry. The large working quarry has a central pit in its floor. On the western face of this depression, ashes show ripple-marks and mud-cracks. This suggests deposition of the volcanic ash in a sea. On the other hand it could have been in a crater lake. The volcanic rocks here at Treffgarne have been quarried for road metal (road chippings). The roughness of each fragment holds them together when they are laid on road foundations. The nearby Precambrian rhyolites are harder but the rock is slippery and these rhyolite fragments would slide against one another. This makes the rhyolite unsuitable for roadstone.

These Precambrian rhyolite lavas and ashes form the outstand-

FIG 16

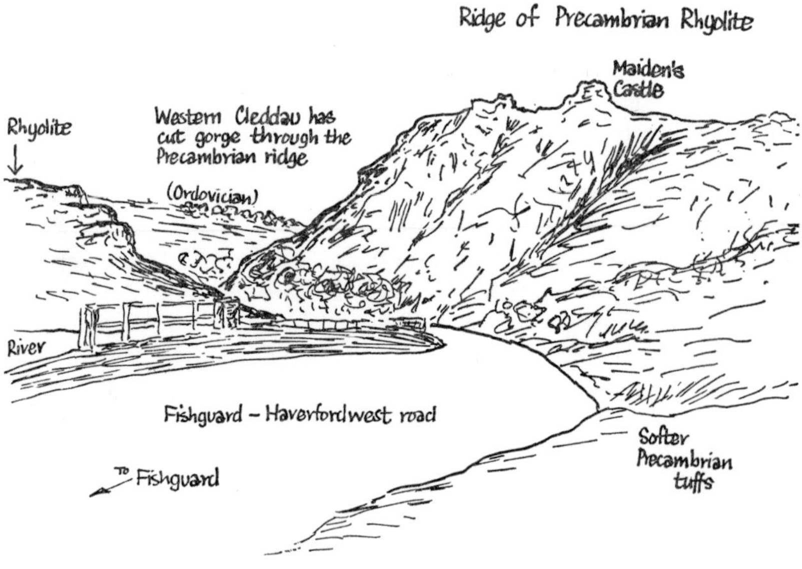

Ridge of Precambrian Rhyolite

Maiden's Castle

Rhyolite

Western Cleddau has cut gorge through the Precambrian ridge

(Ordovician)

River

Fishguard – Haverfordwest road

To Fishguard

Softer Precambrian tuffs

ing crags of Maiden's Castle (an Iron Age camp), visible to the north of the Treffgarne quarries. The ridge of tough, Precambrian rhyolite lavas extends westwards to Roch (near Newgale) and reaches heights of nearly 600 feet in places. North of Treffgarne the River Cleddau has had to cut a steep-sided gorge through the ancient lavas (Figure 16). On the roadside cutting in the gorge, the rhyolite can be seen to be an extremely hard, siliceous rock. On the hilltop above the gorge the rhyolite crags have a very gnarled outline. They can be seen from a considerable distance. The finer grained ashes are seen in the brook to the north of Maiden's Castle and in a lane just north of the roadside farm. These ashes are glossy and very soapy to the touch. The Precambrian volcanics are faulted to the north and south against the Ordovician. This upfaulted mass of older rocks thus forms a 'horst' and the whole of the Cambrian succession is therefore cut out by the faults. Some of this Cambrian sequence can, however, be seen around the village of Ford, one mile upstream from Maiden's Castle.

FISHGUARD

Fishguard is the port for the steamer to Rosslare, Ireland. The town is made up really of three parts—Goodwick Harbour, the main town, and Lower Town at the head of the Gwaun ria. Lower Town will be known to filmgoers as the setting for the film version of Dylan Thomas's 'Under Milk Wood'. The wide Fishguard Bay is carved into the softer Middle and Upper Ordovician shales and flags preserved in the centre of the Goodwick Syncline (Figure 17). The older Ordovician of the syncline's limbs is composed of very thick volcanic rocks (the Fishguard Volcanic Series) and these very much harder rocks—rhyolite lavas and ashes, agglomerates and basic pillow lavas—form the high rugged ground to the west and east of Fishguard Bay. To appreciate the differing resistance to erosion of the soft Goodwick Shales and the Fishguard volcanics, one has only to take a car (or better still, a coach) up the steep hairpin road from Goodwick towards Llanwnda!

The Fishguard Volcanic Series reaches a thickness of over 3,500 feet on the rocky peninsula of Strumble. The volcanic lavas and ashes were erupted during the portion of Ordovician

FIG 17

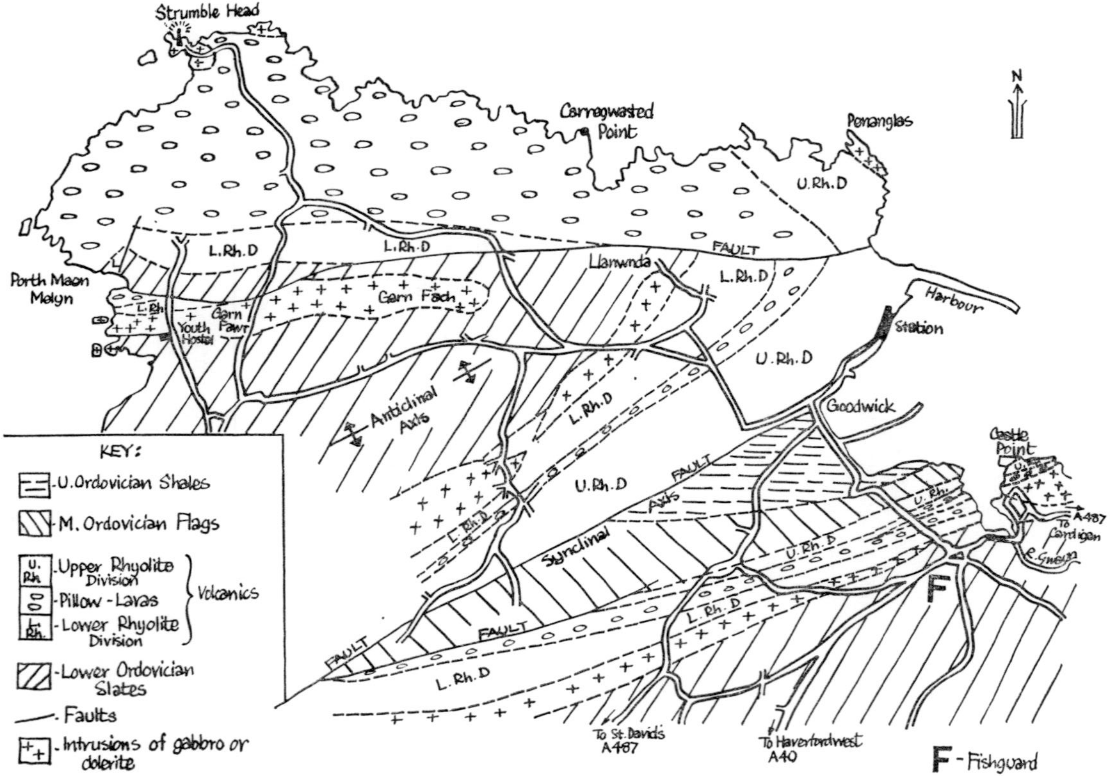
Strumble Head
Carregwasted Point
Penanglas
N
Porth Maen Melyn
L.Rh.D
L.Rh.D
FAULT
Llanwnda
L.Rh.D
U.Rh.D
Harbour
Station
Garn Fach
Garn Fawr
Youth Hostel
U.Rh.D
Anticlinal Axis
L.Rh.D
Goodwick
Castle Point
U.Rh.D
FAULT
Synclinal Axis
U.Rh.D
A487
To Cardigan
R. Gwaun
L.Rh.D
FAULT
FAULT
L.Rh.D
To St. David's A487
To Haverfordwest A40
KEY:
U.Ordovician Shales
M.Ordovician Flags
U. Rh. Upper Rhyolite Division
Pillow - Lavas
L. Rh. Lower Rhyolite Division
Volcanics
Lower Ordovician Slates
Faults
Intrusions of gabbro or dolerite
F - Fishguard

time known as the Llanvirn (the time of the 'tuning fork' graptolites). That the vents were situated to the west of Fishguard is
suggested by the maximum thickness of the volcanics in the
Strumble area, as opposed to the much thinner accumulations
around Lower Town. Mr T. M. Thomas has suggested that a
remnant of one of the volcanic pipes is preserved as a rocky
knoll near Caer-lem Farm (one and a half miles south-south-east
of Strumble Head). Another vent was located at the village of
Goodwick. The acid rhyolites and tuffs were probably associated
with a large number of subaerial volcanoes (Figure 18). Two
separate periods of rhyolitic extrusion in the Fishguard district
were separated by a great extrusion of submarine lava on to the
sea floor, mainly to the west of Fishguard. This lava was of spilite
(a sodic basalt rock) and congealed into great pillow-like heaps.
The shores of ancient volcanic islands may be indicated by
intraformational sands and pebby ashes on the eastern side of
Strumble Head.

The great massive headland west of Fishguard Bay is known
as the Pen Caer Peninsula. Its most north-westerly projection is
Strumble Head, with its visible white lighthouse. The main
structure of the peninsula is a NE-SW aligned anticline (the
Llanwnda Anticline). This upfold, and the Goodwick Syncline
to the south-east, has a plunge downwards to the north-east.
Note in Figure 17 how the rock formations 'V' around these two
folds, because of the plunging or 'pitching' character of the folds.
Notice also how long intrusions (probably sills) of dolerite and
gabbro also conform to the V-like pattern. This shows that these
intrusions occurred before the rocks were folded.

The best exposures of the pillow lavas occur near the car park
for the Strumble lighthouse. By walking a little way back eastwards along the narrow road, a disused look-out station (Grid
Ref 895413) can be seen. One can make one's way down an easy
slope to the east of this building and on to the rocky foreshore.
Here and in the nearby face of an inlet, excellent exposures of
pillows can be seen (Figure 19). Dark ashy shales, interbedded
with the pillow lavas, are seen to be baked.

Another locality worth visiting on the Pen Caer Peninsula is
the little inlet called Porth Maen Melyn ('Yellow Stone Gate'),
on the west coast. This is not far from the Youth Hostel. That
hostel is at the western end of the great rocky ridge of Garn

FIG 18

Falling Ash

Volcanoes near Builth

Lava

The Ordovician Sea over Pembrokeshire

FIG 19

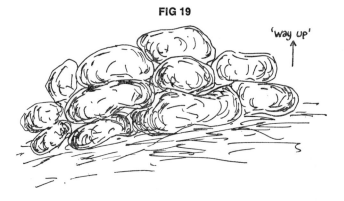

'way up'

Fawr (formed of dolerite). The numerous headlands around this hostel are formed of igneous rocks whereas the narrow inlets are either along zones of weaker slates or along faults. The south side of Porth Maen Melyn is formed of pillow lavas whereas the northern cliffs of the inlet display excellent exposures of rhyolite and of course, rhyolitic agglomerates. In between is the deep indent formed in the weaker, fault-riddled Llanvirn slates. The bay is an excellent example of the control on topographical and coastal outline by the geology and structure. The striking rocky ridge of dolerite forming Garn Fawr and its eastern neighbours is a further pointer to the close relationship between geology and land surface in this Fishguard area.

Before leaving this western side of Fishguard, mention must be made of the last invasion of Britain, in 1797, by the French on the coast of the Pen Caer Peninsula. The landing took place on Carregwastad Point (Figure 17) and the French occupied many of the farmhouses. The French had in fact sailed for Ireland but landed in Pembrokeshire by mistake. They tried to raise the local population to their cause but soon surrendered to Lord Cawdor and the local militia (later to become the Pembroke Yeomanry). A much-told version of this invasion suggests that the French panicked when they saw Jemima Nicholas and her woman friends parading their red Welsh shawls on the cliff tops, the foreigners mistaking the shawls for the red uniforms of British soldiers. There seems to be some doubt about this well-loved version.

THE PRESCELLY HILLS AND STONEHENGE

Fishguard's River Gwaun rises on the western portion of an E-W range of hills known as Prescelly. These hills reach heights of between 1,000 and 1,750 feet OD and form a striking feature of the topography on account of their barren but graceful outline and their abrupt emergence from the much lower plateaux of Pembrokeshire. Along the hill crests ran the most ancient of western ways—Ffordd Fflemming. The hills abound in stone circles, standing stones, tumuli, megaliths, etc. The Prescelly Hills might well be considered to be the Western Piccadilly of Neolithic to Bronze Age times.

The rocks of the Prescelly hills are of Ordovician age and fall generally into three main types—Ordovician slates, rhyolitic lavas, ashes and tuffs (with some trachytes) and intrusions of quartz-dolerite. The volcanic rocks are limited to the Lower Ordovician and in part form an easterly continuation of the Fishguard Volcanic Series. The intrusions affect these lower Ordovician rocks but do not appear to be intruded into Middle and Upper Ordovician strata. This could mean that the intrusions

date from Lower Ordovician times and were associated with the
volcanic outbursts. The higher hills are associated with the long
tongues of dolerite intruded into Llanvirn slates. Carn Meini and
Cerrig Marchogion are good examples. Foel Trigarn and Carn
Alw are, however, formed of the flinty rhyolites. The overall
geological structure of the hills involves two eastward-pitching
anticlines, separated by a syncline. There is much overthrusting
to the south, that is on to the eastward continuation of the
St David's Anticline.

It seems a far cry from North Pembrokeshire to Salisbury Plain
but this is where the story has to be now taken up. Stonehenge
is the finest Megalithic circle in Britain. It dates back to at least
the Bronze Age and was rebuilt three times before Christ was
born. The stones of Stonehenge can be subdivided into two main
groups and kinds. The large stones of the Outer Circle are
'sarsens', whereas those of the Inner Circle, the horse-shoe and
the altar stone, are familiarly referred to as the 'bluestones'
(Figure 20). The sarsens are silicified quartzose sandstones that
may have once formed part of a deposit of Eocene 'Bagshot
Sands' in southern England. Dean Buckland proved the identity
of these Sarsens with blocks scattered all over the Salisbury Plain
and commented on the abundance of these blocks in regions like
Savernake Forest, Hungerford and near Marlborough. Buckland's
views were confirmed by later findings of Conybeare, Prestwich,
Philips and Ramsey. Most of the large sarsens were probably
collected by the builders of Stonehenge from boulders within the
immediate neighbourhood of the circle site. In this way Stonehenge compares with its counterpart at Carnac in France.

The unique feature of Stonehenge is the considerable number
of the 'foreign' bluestones. These can be further subdivided into
three categories: (a) Dolerites, (b) Rhyolites and (c) the micaceous sandstone of the 'altar stone'. The last-named rock is a pale
green sandstone, very similar to the Senni Beds of the Old Red
Sandstone in South Wales or to the Cosheston Beds of the same
system in mid-Pembrokeshire (near Milford Haven). Four of the
Stonehenge stones are rhyolites. The rocks are flinty and dark
grey with a flow-banding of narrow, frequent, parallel lines. The
felspar in these rhyolites is rich in soda. Spherulitic growths due
to incipient crystallisation and curved (perlitic) cracks are further
characteristics of the four Stonehenge rhyolite blocks. All these

FIG 20

Stonehenge

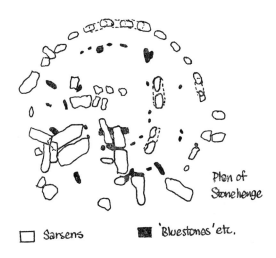

Plan of
Stonehenge

☐ Sarsens   ■ 'Bluestones' etc.

features can be matched in the rhyolites of Foel Trigarn and
Carn Alw in the Prescellies, as seen in the descriptions by
Parkinson in 1897. Lastly, there are no less than twenty-nine
dolerite stones at Stonehenge. These are moderately coarsely
crystalline rocks of blue-green to greenish-grey colour. Two
features make it fairly certain that all the dolerite stones came
from the same source. First, the dominant felspar is rich in soda.
Secondly, the rocks are spotted with white or pinkish felspathic
spots of all sizes. These spots are composed of irregular crystals
and crystal groups of felspar. The presence of these two features
in the dolerites of the Prescelly Hills makes it virtually certain,
as shown by H. H. Thomas in 1923, that the twenty-nine blue-

stones of Stonehenge came originally from the Prescelly Hills.

The 64,000 dollar question now, of course, is 'How did these thirty-four foreign stones get all the way to Stonehenge?' There are two possible answers: (1) carriage by ice; (2) carriage by man. There might of course even be a third answer: carriage by man of material that had been carried some of the way by ice. The generally accepted view is that man carried the stones all the way from Pembrokeshire. Professor Atkinson has suggested the route, via the Milford Haven, the skirting of the northern coasts of the Bristol Channel, by raft up the Bristol Avon and thence overland, on rollers, to Salisbury Plain. The mode of transport has been proved by being performed today. This theory fits in with the picking up of a boulder from the Milford Haven Cosheston Beds on the way. H. H. Thomas had earlier favoured human transport. One might ask: 'Why did man go to all this trouble, bringing these stones all the way from Pembrokeshire, 190 miles away, and with blocks up to five tons in weight to manoeuvre?' It may be significant here that many stone circles existed in North Pembrokeshire and that when it was decided to build a great temple at Stonehenge, those Pembrokeshire stones might well have had a pride of place, with some mystic or religious significance.

Recently, G. A. Kellaway has reintroduced the glacial theory for the carriage of the foreign stones to Salisbury Plain. He has drawn attention to recent finds of boulder clays and of chalk fragments well into Somerset and as far as Holwell in the eastern Mendips. This locality is no more than twenty-five miles from Stonehenge. Kellaway has drawn attention also to the large erratic boulders found in south-west England and to the possibility that earlier ice sheets of the Pleistocene Ice Age may have brought western erratics into Salisbury Plain. Moreover, he has thrown doubt on the earlier views about the local derivation of the sarsen blocks, suggesting that they could have been brought by ice from submarine Tertiary sandstone deposits even in the English Channel. He also draws attention to the *other* rocks found at Stonehenge during excavations. These include lumps of slate, quartzite, flagstone, sheared volcanic ash and greywacke. Why carry all this strange mixture to Stonehenge when there was an ample supply of good dolerite on the Prescelly Hills? Kellaway puts forward the alternative explanation that this rock mixture was more feasibly a glacial-carried mixture. If Kellaway is right, not only is the Stonehenge mystery solved but a new picture emerges of extensive earlier ice extensions into southern England, far beyond the Bristol Channel. Neolithic and Bronze Age man might still have been sentimentally attached to the Prescelly bluestones, whether he was erecting temples and circles in North Pembrokeshire or on Salisbury Plain. Ice sheets, however, helped him in the building of the Stonehenge temple.

CHAPTER 3

South Pembrokeshire

The South Pembrokeshire coastline is well known to the thousands of visitors who come to this attractive part of South Wales each summer. There are beautiful sandy coves, imposing cliffs of red and light grey colours, islands such as Skomer and Caldy. It is said that Pembrokeshire is a 'little England beyond Wales', but one could equally say that the county is a 'Devon beyond England'. To the ornithologist, the geologist and the historian, the southern half of Pembrokeshire has particular attractions. There are many historic castles. The imposing castle at Carew (near Pembroke), built by Norman conquerors in the thirteenth century, became the home of Sir Rhys ap Thomas of Dynevor, who built the Great Hall there. Further portions were added in the late sixteenth century by Sir John Perrott, said to be an illegitimate son of Henry VIII. One of the earliest castles built by the Normans in South Wales is that at Manorbier, dating back to around 1090. This castle was supplied by the sea route. The famous writer on Wales, Giraldus Cambrensis, was born here in 1146. The magnificent castle at Pembroke is well known to all Pembrokeshire visitors. Much visited also is the Cistercian monastery on Caldy Island, off the Tenby coast. The modern building is on the site of a medieval structure and there are two ruined churches nearby. St Govan's Chapel, a tiny structure set in a cleft in the limestone cliff top of the south coast, was the tiny cell of the Celtic saint, Govan. The present chapel is thirteenth century. The holy well nearby is now dry.

Geologically, South Pembrokeshire has much to offer. The powerful Hercynian (Armorican) earth movements, which occurred at the close of the Carboniferous period, intensely buckled the rocks of South Pembrokeshire. The limbs of these folds are frequently tilted into a vertical position (see Figure 21).

Anticline and syncline follow one another across the region (Figure 22). These folds were subsequently bevelled down by forces of erosion in Permian and Triassic times and further bevelling was carried out by the sea in late Cenozoic times. The result of this erosion now is to reveal a tremendous range of strata from the Ordovician to the Coal Measures of the Carboniferous. The deeply eroded cores of the upfolds reveal the Ordovician and Silurian rocks whilst in the centres of the synclines the Carboniferous strata are preserved. Powerful fractures break the continuity of the folds and there are many important thrusts, including the lengthy Ritec Fault which runs from Tenby on the east coast to the Dale Peninsula in the west. Active

FIG 21

"Three Chimneys" Marloes Bay: Pembs.

FIG 22

Broad Haven: Pembs:

erosion along this fracture, followed by recent drowning, has produced the spectacular Milford Haven.

There is so much to see, geologically, in South Pembrokeshire that only a few of the more important localities are described here. An attempt is made to describe them from the point of view of an *ascending* succession. As the Ordovician exposures at Freshwater West and Freshwater East are not now visible, the account begins with the Silurian rocks of Marloes Sands.

MARLOES SANDS

This beautiful sandy bay is situated on the south side of the

Wooltack-Skomer promontory and about one and a half miles south-west of the village of Marloes. Cars and small coaches can proceed along the narrow road which leads in the direction of the western end of Marloes Sands, halting at the car park (Grid Ref. 780082). From near here a narrow track, known as Sandy Lane, leads to the western portion of the Sands. Figure 23 shows

FIG 23

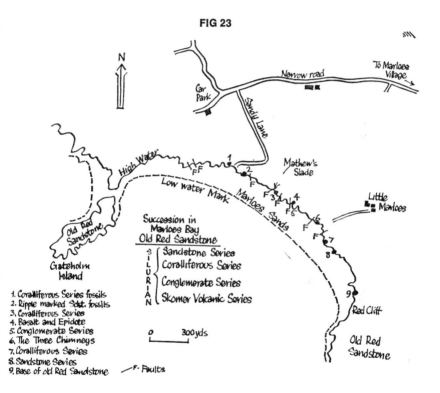

Succession in
Marloes Bay
Old Red Sandstone

S ⎰ Sandstone Series
I ⎱ Coralliferous Series
L
U ⎰ Conglomerate Series
R
I
A
N ⎱ Skomer Volcanic Series

1. Coralliferous Series fossils
2. Ripple marked Sdst. fossils
3. Coralliferous Series
4. Basalt and Epidote
5. Conglomerate Series
6. The Three Chimneys
7. Coralliferous Series
8. Sandstone Series
9. Base of old Red Sandstone

0   300yds

—F- Faults

the points of geological interest in this bay. The west coast of the embayment is dominated by the Old Red Sandstone island of Gateholm. Further out to sea is Skokholm, the bird sanctuary.

The Silurian succession at Marloes Sands is divisible into three parts:

    (3) The Sandstone Series   1,000 feet
    (2) The Coralliferous Series   300 feet
    (1) The Skomer Group   300 feet seen
        (base not visible)

The Skomer Group is best developed on that island where there are thick sequences of volcanic basalts, rhyolites and tuffs, together with conglomerates and sandstones. This great volcanic outburst of Skomer was previously ascribed to early Ordovician times but recently it has been shown that these Skomer volcanics pass upwards and grade laterally into sediments containing Lower Silurian (Llandovery) fossils. The volcanic outbursts probably took place near or even on land. The top portion of the Skomer Group is largely of sedimentary type, with conglomerates, sandstones and shaley mudstones. These beds are followed by the grey Coralliferous Series, consisting of calcareous shales, mudstones and siltstones with calcareous lenses. Some of the beds weather as 'rottenstone', brown in colour and showing the fossils more clearly. The Sandstone Series comprises alternations of grey, brown and yellow-green sandstones with siltstones, shales and rottenstones. Most of the strata in Marloes Bay dip steeply southwards and in places are vertical. In this Marloes area there is no break between the Silurian and Devonian systems and an arbitrary boundary is taken at the first appearance of red beds (at **9** in Figure 23).

The first rocks seen at the beach end of Sandy Lane are rottenstones belonging to the top of the Coralliferous Series. Brachiopods such as *Atrypa* and *Eospirifer* can be found. From here and around the bluff to the east as far as the broad grassy break in the cliffs (Mathew's Slade), the overlying Sandstone Series is well exposed. A number of faults can be clearly seen and excellent ripple-marks are visible on some of the bedding planes (**2**). At Mathew's Slade a fault wedge brings back the Coralliferous beds (**3**). A hundred yards south-east of the foot of the beach path in the slade, the Skomer Group is brought up by faulting and about 75 feet of basalt is seen. There are two basalt flows, the lower one having an uneven red-weathered top. The basalts are vesicular, that is full of gas holes. Some of the holes contain a yellow epidote, as also do veins which cut the lava flows. The brecciated and red-weathering tops of the lavas suggest strongly that the volcanic extrusion took place in a non-marine environment. Basalt flows in tropical climates today weather in a similar way. The lowest basalt can be seen, in a deep cleft, to lie on top of yellow-grey felspathic sandstones, possibly a beach deposit. Above the basalts come coarse conglomerates, ashy mudstones and a thick agglomerate, the latter forming a stack on the foreshore. Large specimens of shining black *Lingula* can be found in some of the ashy mudstones. These lingulids occur on bedding planes very near to the three well-marked bands of sandstone, standing vertical and known as the Three Chimneys. (Figure 21). The succeeding beds yield *Lingula, Leptaena, Camarotoechia, Tentaculites* and crinoid stem fragments.

The overlying Coralliferous Series comes on in the cliffs just beyond a fault. The junction is discordant with a difference of dip of about 15 degrees. The roof of a nearby small cave yields abundant fossils, including *Eocoelia* (a brachiopod) and the coral *Palaeocyclus*. Further corals, especially the tabulate coral *Favosites*, can be found in the continuing outcrops of the Coralliferous Series. Numerous faults occur in this vicinity and there are excellent offsetting veins of quartz suggesting a tensional stretching of the rocks, probably at the time of formation of the faults. The pattern of the tension gashes is reminiscent of the cracks in the ceilings of a newly constructed house.

The junction of the Coralliferous and Sandstone Series is obscured by fallen blocks. The Sandstone Series is rather barren but the few fossil bands that do occur show that the Pembrokeshire Silurian faunas have more in common with those in Nova Scotia than with the Silurian of Shropshire (the type area of these Upper Silurian strata). One should remember, of course, that in Silurian times North America was very near to Britain. A drifting westwards of the American area began in Mesozoic times and still continues today. The region of split is in mid-Atlantic where new basaltic material is welling up from below to occupy the ever-widening gap between the drifting plates.

The steep dip of the beds here is somewhat seawards of the line of the coast and this factor, together with the presence of shaley horizons within the sandstones, has led to some landsliding of the strata down the cliffs. The resultant deformation of the beds is clearly seen at the base of the cliff where some blocks have virtually rotated into an overturned position. A return to normal dips takes place in the next minor headland where the beds show channelling and contain worm burrows. Beyond this headland is a cove (to be entered only at low tide) which shows the gradual alternation of grey with red beds. The

Geological Survey in 1916 inscribed a benchmark-type arrow here to indicate their selection for the base of the Devonian System.

FRESHWATER EAST

This bay is reached from Tenby by taking the A4139 and turning off southwards at Lamphey. A steep minor road descends to the bay from Freshwater East village. The most interesting section is on the south side of the bay. Ordovician slates were once visible in the core of the Freshwater East anticline, but sand has subsequently hidden them. The junction of the Devonian with the thin underlying Silurian sandstones, sandy shales and rotten-stones can however be clearly seen near the south-west corner of the bay. The Silurian rocks, brown in weathered colour, are very fossiliferous in certain bands, especially the rotten layers which yield the trilobite *Homalonotus*, the gastropod *Holopella*, and the brachiopods *Marklandella*, *Atrypa* and *Lingula*. The basal beds of the Devonian are red and greyish-green, coarse conglomerates with some very large pebbles of quartzitic rocks (Figure 24). Above these pebbly beds come red standstones and

FIG 24

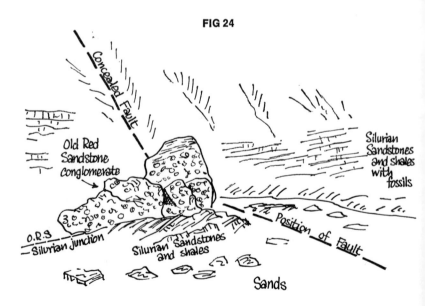

FIG 25

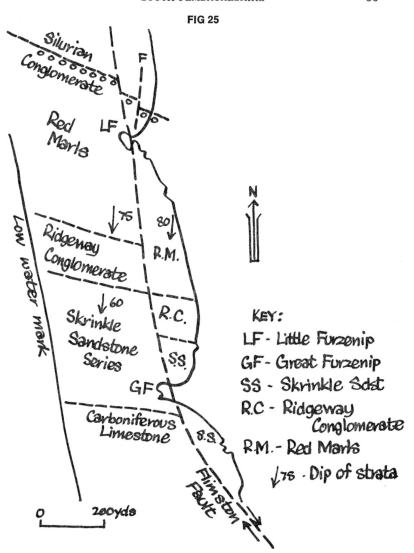

KEY:

LF - Little Furzenip
GF - Great Furzenip
SS - Skrinkle Sdst
R.C - Ridgeway
       Conglomerate
R.M. - Red Marls
↓75 - Dip of strata

shales. There are several faults in this cliff section. A big one
occurs to the western side of the conglomerate crags.

FRESHWATER WEST

This bay is on the west coast and like its eastern counterpart has an anticlinal structure, with Lower Palaeozoic strata being exposed in the core. Once again it is difficult to detect the Ordovician slates once seen here. The highest Silurian is still visible, however, and consists of siltstones and sandstones with thin calcareous layers yielding brachiopods, lamellibranchs and gastropods. Once more the basal layers of the Old Red Sandstone are coarse pebbly conglomerates.

The main feature of this locality (Figure 25), however, is the excellent section of Old Red Sandstone displaced in the cliffs and on the rocky foreshore. From the B4319 road adjacent to the headland known as Little Furzenip, a good view can be obtained of the bay, and the line of the great Flimston tear fault can be clearly seen. The fault gap in the rock platform is visible and the fault can be seen to isolate the Little Furzenip stack from the headland. This fault has suffered a lateral shift of almost 200 yards, the western side having moved northwards relative to the eastern block. There is a possibility that this fault could continue across the Bristol Channel to become the famous Sticklepath Fault of Devon. That fault too is a tear fault with a northward slip of its western side.

Above the basal conglomerate of the Devonian come about 1,300 feet of red marls, sandstones and concretionary limestones, called 'cornstones'. Numerous green bands also occur, and a cyclic repetition of the various rock types can be detected, especially in the middle beds of the Red Marls. The sediments were probably deposited in a fluvial environment with numerous meandering streams. Sandstone plugs or pipes up to 6 inches across may represent 'sand volcanoes' formed through erosion by rising ground-water.

(Before proceeding further south along the bay, permission should have been sought from the Military Camp at Castlemartin, whose firing range embraces the Great Furzenip headland.)

The Red Marls are followed by a very different succession of coarse red conglomerates, interbedded with sandstones and siltstones. This formation is the Ridgeway Conglomerate. Studies have shown that this pebbly debris was brought in by rivers from the south. The pebbles, some of which contain fossils, show that the southern area that was being eroded was one of Cambrian and Ordovician rocks. The exact Devonian position of the Ridgeway Conglomerate is unknown; it could, at least in part, be of Middle Devonian age—rocks otherwise unknown in South Wales.

The conglomerate is succeeded by the Skrinkle Series, made up of red sandstones, siltstones, white and buff quartzites, conglomerates and, near the top, marine limestones and shales. The age of the Series (Upper Devonian) was proved by the finding of scales of the fish *Holoptychius* (a crossopterygian ganoid fish, allied to the Coelocanth) in the lowest beds in 1921. Part of the Skrinkle Series is of marine origin, showing that the southern sea (over Devon and Cornwall) was making attempts to invade South Wales.

To see the junction of the Skrinkle Series with the Carboniferous, it is advisable to visit West Angle Bay (four miles to the north-west). On the north side of the bay it is possible, tides permitting, to see the passage up from the Skrinkle Series into the Lower Limestone Shales of the Carboniferous. Spectacular minor folds occur in the Lower Limestone Shales on the north side of West Angle Bay and there is some thrusting too.

STACKPOLE QUAY

This cove is situated some four miles south of Pembroke and on the east coast of southernmost Pembrokeshire. The area displays some of the structural elements of South Pembrokeshire —sharp folds trending E-W and dislocated by a main cross fault which displaces the axes of the folds (Figure 26). A good sequence of the Carboniferous Limestone is exposed and the junction with the underlying Old Red Sandstone is seen in another cove to the north. The main quay is eroded into an anticline involving soft calcareous mudstones in the middle part of the Limestone succession. This is a good example of the greater erosion along the crest of an upfold. This anticlinal axis is displaced by a hundred yards to the north on the western side of the main Stackpole Quay Fault. Erosion along this fault is responsible for the long inner recess of the cove. This fracture can be clearly seen in the northernmost recess of the quay. The fault plane has a steep dip and almost horizontal slickensides (polished grooves) can be seen on the fault face. Mudstones on the western side can be seen to be faulted against crinoidal limestones on the east. Those same limestones are seen in the large quarry (1) to the north-west (on the western side of the fault). Directly seawards of that quarry, a sharp syncline is visible on the small island. The small inlet 350 yards north of that island is excavated into the softer Lower Limestone Shales. The shales are much affected by thrusting, but the junction with the under-lying Skrinkle beds of the Old Red Sandstone is seen beyond the northernmost thrust. A grey limestone, 1 foot thick and con-taining crinoid debris, pink quartz granules and black phosphatic nodules, rests on 16 feet of red concretionary siltstone with green mottling at the top. This colour change is the Devonian-Carboniferous junction. The Upper Devonian brachiopod *Cyrtos-pirifer verneuili* (the 'Delabole Butterfly' of the Delabole slate quarries in Cornwall) occurs in shelly sandstones some twenty

FIG 26

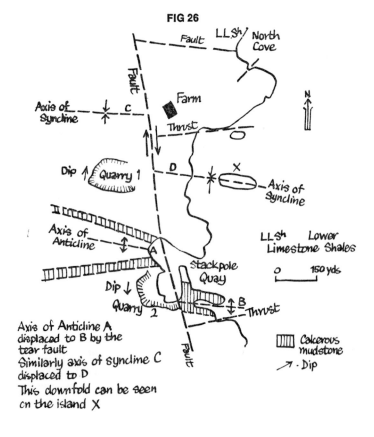

FIG 27

"Ladies Cave Anticline" Saundersfoot Pembs.

to thirty feet further down the Devonian sequence. The headland (First Point) on the north side of this North Cove is formed
of tougher conglomerates in the Skrinkle Series. Northwards
these pebbly beds pass down into quartzites, but then these
lowest Skrinkle beds rest directly on the Red Marls. Thus the
Ridgeway Conglomerate of Freshwater West is missing here and
must have been eroded away before the Skrinkle Series was
deposited.

TENBY

The hill promontory of Castle Hill and the nearby St Catherine's Island (with its old fort) separates the two stretches of sands—North Sands and South Sands—for which the resort of Tenby is renowned. These two sand stretches are also noted for the geological sections seen in their facing cliffs. At South Sands there is a good section in the Carboniferous Limestone whilst at North Sands there is a good section in part of the Millstone Grit (Figure 28). Between the two sections trends the complex Ritec Fault. This is a complicated zone of E-W fractures, involving the northward pushing of older beds on the south over younger strata to the north. One group of thrusts runs along the south flank of Castle Hill and through to the harbour. The limestones on the southern edge of Castle Hill can be seen to be much disturbed and riddled with red-stained calcite veins. Another fault divides St Catherine's Island from the South Sands section. One of the main Ritec fractures runs close to the Barrel-post Rock (in the southern end of North Sands). This fault member thrusts the higher limestones against beds in the middle of the Millstone Grit.

The Carboniferous Limestone of the South Sands section is folded into a major anticline, the axis of which occurs almost in line with the southern tip of St Catherine's Island. The core of the upfold exposes dolomites and oolites overlain by softer calcareous mudstones which are seen in marked narrow inlets in the cliff. Above these mudstones, on the south flank of the anticline, come further dolomitic limestones and then light grey oolitic limestones with *Seminula* and *Lithostrotion* colonies.

The North Sands cliffs expose folded and faulted Millstone Grit. Despite the folds, however, the strata dip predominantly towards the south. Yet despite this southward dip the strata generally become *younger* as one goes northwards from Tenby. The reason for this apparent mystery is that *blocks* of southward dipping strata are consistently thrust upwards and northwards over other blocks. As a result, progressively younger blocks occur in a northward direction. Many of the thrusts are difficult to locate because the fault planes are often parallel to the dip of the beds. Fossils are then the only clue to the age of

FIG 28

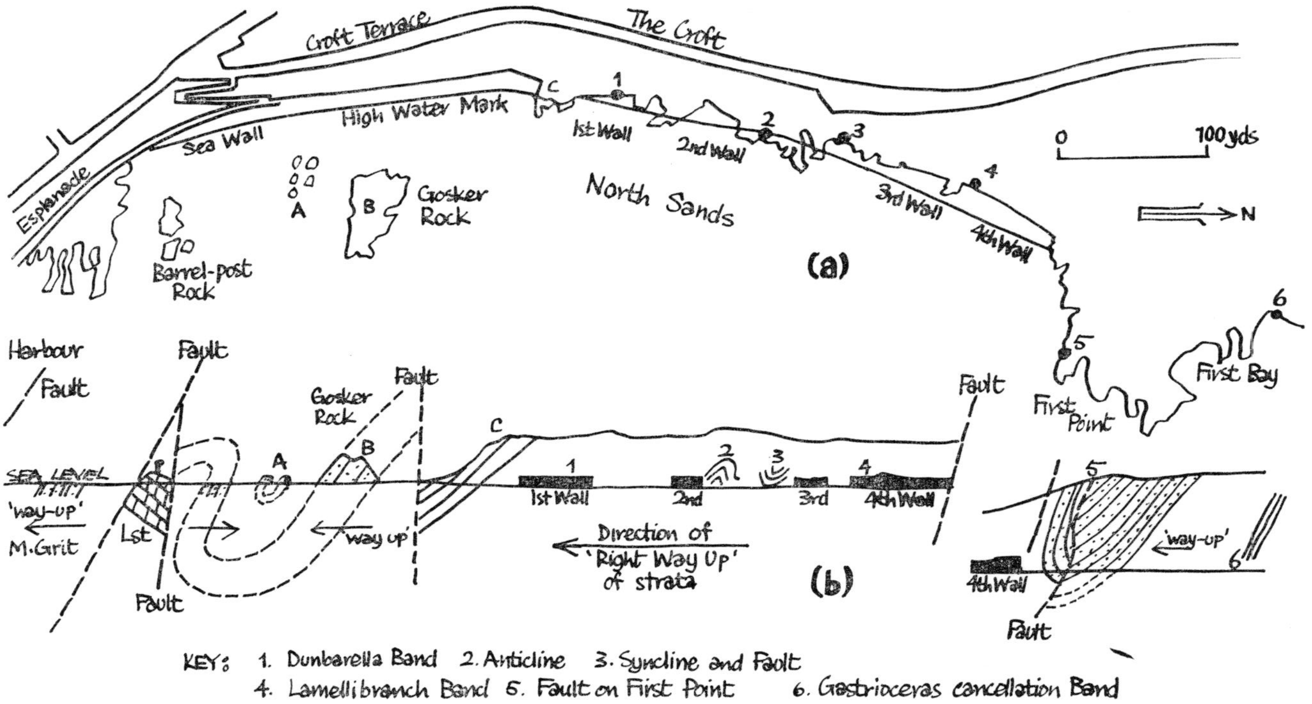

Croft Terrace
The Croft
Esplanade
Sea Wall
High Water Mark
C
1
1st Wall
2
2nd Wall
3
3rd Wall
4
4th Wall
0
100 yds
N
North Sands
A
B
Gosker Rock
Barrel-post Rock
(a)
5
First Point
6
First Bay
Harbour Fault
Fault
Gosker Rock
Fault
Fault
SEA LEVEL
'way-up'
M.Grit
Lst
Fault
way up'
1st Wall
2nd
3rd
4th Wall
Direction of 'Right Way Up' of strata
(b)
'way-up'
4th Wall
Fault
KEY: 1. Dunbarella Band 2. Anticline 3. Syncline and Fault
4. Lamellibranch Band 5. Fault on First Point 6. Gastrioceras cancellatum Band

a particular block of strata. Particularly useful are the coiled Cephalopods known as Goniatites. Bands with the goniatite *Anthracoceras* occur just north of the anticline (**2**) between the second and third walls (see Figure 28). Much higher goniatite horizons can be found in the shales beyond First Point (**6**). Other fossil bands on the North Sands section contain lamellibranchs— *Dunbarella* behind the first wall and *Sanguinolites* (a small Lingula-like lamellibranch) on either flank of the anticline mentioned above. The sandstone beds on Goskar Rock the rocky island in the middle of North Sands) were once thought to be upside down. Lumpy projections below the base of sandstone beds ('load casts') show, however, that the beds are the right way up. The Goskar Rock sandstone is probably the same as the sandstone (**C**) just south of the first wall. A fault must therefore occur between these two sandstones. A rather undulating fault plane (**5**) traverses the south side of the First Point headland, bringing almost overturned beds on the south side against southward-dipping strata on the north. The fault must virtually cut through the centre of an overturned syncline.

SAUNDERSFOOT

This growing boating resort has fine cliffs of the Coal Measures in its vicinity. The cliffs south of the harbour are in the Lower Coal Measures whilst those north of the resort are in the Middle Coal Measures. A powerful line of thrusting must separate the two sections but the fault belt is concealed beneath the town. The best section is undoubtedly that between the harbour and Monkstone Point. These cliffs reveal vividly the powerful crumpling of the Coal Measures by the Hercynian earth movements and there are many good examples of thrusts.

The rock wall forming the south side of the harbour is the almost vertical northern limb of an anticline. The beds belong to the Communis Zone of the Lower Coal Measures. The large rounded masses of reddish-brown colour are nodules of clay-ironstone. These ironstones were once in demand in South Wales, in the early days of iron smelting, and were much mined and dug along the North Crop of the main coalfield, especially at Glynneath, Dowlais and Ebbw Vale. The gentler southern limb of the harbour anticline is seen beyond the steps down on to the beach, south of the harbour. A few yards further south along these cliffs some very complex structures occur in the shales and thin sandstone ribs with several ragged faults and small scale folding. The next projecting minor headland is a fine example of a thrust overfold, best viewed from the direction of the sea. A coal seam overlies a thick fireclay here but the folding and thrusting has virtually sheared out the coal or crushed it into 'small coal' or 'culm'.

A broader gap then occurs beyond which is a major projecting cliff. This cliff forms the northern limb of the most famous anticline in South Wales and one which has been photographed on countless occasions, occurring in many famous geological textbooks. Once again the best view of the fold is from the sea side. The fold is the Ladies Cave Anticline (Figure 27) and again the beds affected are in the Communis Zone of the Coal Measures. Note the much eroded remnants of the fold on its seaward side.

Many thrusts, with low angles of dip, occur in the next stretch of cliff. Some of these faults dip at the same angle as the strata affected and detailed fossil identification of horizon is necessary in detecting their presence. A major overfold can be made out in massive sandstones occurring high in the cliffs. Climbing over a low rock wall and on to ledges at the foot of the next cliff wall, the sandstones are seen to have a markedly curved base, cutting into the shales beneath. These dark shales are fossiliferous, containing non-marine lamellibranchs (see Figure 32). The forms *Carbonicola crispa* and *Carbonicola pontifex* can be found and show that these beds now belong to the lowest zone of the Coal Measures—the Lenisulcata Zone. This fossil horizon is one of very wide distribution, being found in other parts of the South Wales Coalfield, in the coalfields of Yorkshire and Lancashire and even as far away as Leinster in Ireland and the Ruhr of Germany.

To proceed beyond this point, low tide is necessary and one must watch for its return. Further spectacular folds and faults occur before the Monkstone headland is reached. The sandstones of this promontory are particularly tightly folded.

The cliff sections between Saundersfoot and Wiseman's Bridge display the Modiolaris Zone Coal Measures (above the Communis Zone). At least two major anticlines can be traversed. The nor-

thern fold underlies the hill on which Hean Castle is situated.
A thick coal—the Garland Coal—can be seen on either flank of
this broad upfold. This coal marks the base of the Modiolaris
Zone in Pembrokeshire.

AMROTH

This seaside village lies on the northern extremity of the
Pembrokeshire coalfield along the Carmarthen Bay coast. The
Lower Coal Measures are again well exposed in the cliffs immediately south of the village. The first structures seen here are
overfolded beds with medium-angled thrusts. This belt of disturbance is probably the Pembrokeshire continuation of the Careg
Cennen-Llandyfaelog Disturbance (described on page 76). Immediately south of this disturbed belt, a small recess excavated
parallel to the cliff face shows the remnants of the lowest coal
seam in this coalfield—the Kilgetty Vein—with its underlying
fireclay and sandy ganister forming the southern lip of the recess.
The roof of the coal is a dark shaley mudstone yielding fish scales.
About sixteen feet up the cliff face from the coal is a 'mussel'
band with *Carbonicola pseudorobusta* (Figure 32).

Walking southwards along the cliffs from here, a well-marked
brown layer comes into sight. The layer—it is the Amroth
Limestone—is almost horizontal but is broken by a number of
small faults. This limestone is a non-marine limestone, formed
of the mineral 'ankerite'; the band lies about thirty feet below
the Kilgetty coal. In a little while the limestone dips down to
beach level and can be seen to be full of the non-marine lamellibranch *Anthracosia regularis*. The upper layers of the bed are
riddled with worm burrows. One last reminder along this coast
of the intense Hercynian folding of the beds at the close of the
Carboniferous Period is supplied by the spectacular S-fold, with
a thrust in its midst, seen in the cliffs nearby.

CHAPTER 4

Across the Black Mountains

In this chapter it is proposed to describe a journey across the Black Mountains of Carmarthenshire, a range of hills reaching to between 1,500 to 2,000 feet above sea level. The route, taken in an anticlockwise direction, is shown in Figure 29, which also shows the position of the various stops. For the purpose of this chapter, the route will begin at Pontardawe, in the Swansea Valley, as the lower portion of that valley is better described in chapter 7.

The route has one very unique feature, in that in the course of about fifteen miles, in a N-S direction, one can traverse from Upper Carboniferous rocks through the Devonian and Silurian systems to the middle of the Ordovician. Moreover, in this journey one can collect Coal Measure 'mussels', Silurian brachiopods and Ordovician trilobites. In many parts of the world one could travel by train or air for considerable distances and remain over the same rocks all the time. The route has, in fact, one further advantage to the student of geology and all who love scenery. It displays most vividly the close relation between the 'ups' and 'downs' of the land surface and the nature of the underlying strata. The Black Mountains succession is a remarkable alternation of hard and soft formations and this explains the irregular character of the topography (Figure 30). The hard formations are the Pennant Measures (Upper Coal Measures), the Farewell Rock and Basal Grits (of the Millstone Grit), the Carboniferous Limestone, the upper portion of the Old Red Sandstone (Grey Grits and Brownstones) and lastly the Trichrug Beds of the Higher Silurian. In contrast, formations like the lower half of the Coal Measures, the Middle Shales of the Millstone Grit, the Red Marls (Old Red Sandstone), and the shales, limestones and slates of the Ordovician System are by com-

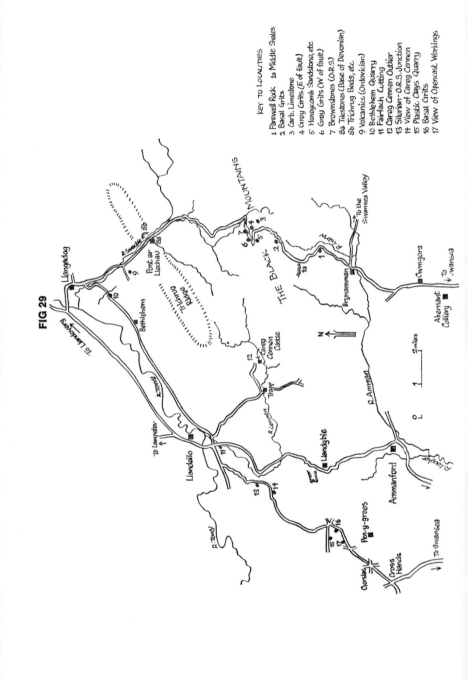

FIG 29

KEY TO LOCALITIES

1 Farewell Rock   1a Middle Shales
2 Basal Grits
3 Carb. Limestone
4 Grey Grits (E of fault)
5 Honeycomb Sandstone, etc
6 Grey Grits (W of fault)
7 Brownstones (O.R.S)
8a Tilestones (Base of Devonian)
8b Trichrug Beds, etc.
9 Volcanics (Ordovician)
10 Bethlehem Quarry
11 Fairfach Cutting
12 Carreg Cennen Outlier
13 Silurian-O.R.S. Junction
14 View of Carreg Cennen
15 Plastic Clays Quarry
16 Basal Grits
17 View of Opencast Workings

FIG 30

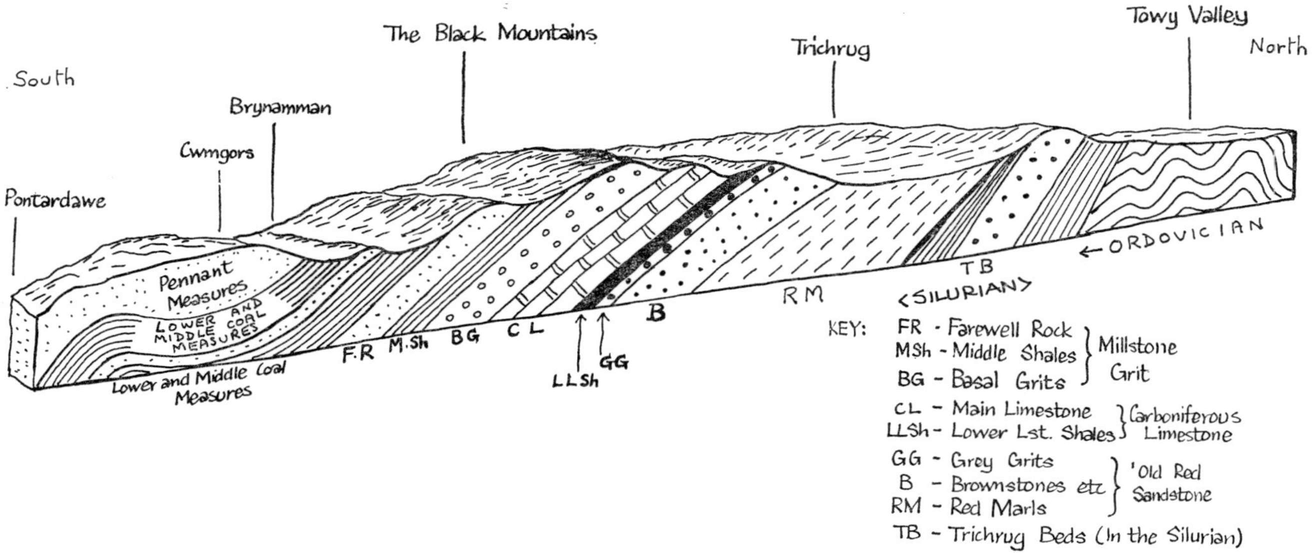

South — Pontardawe — Cwmgors — Brynamman — The Black Mountains — Trichrug — Towy Valley — North

Pennant Measures

LOWER AND MIDDLE COAL MEASURES

Lower and Middle Coal Measures

F.R M.Sh B.G C.L LLSh G.G B RM TB

←ORDOVICIAN→

←SILURIAN→

KEY: FR – Farewell Rock } Millstone
      MSh – Middle Shales }
      BG – Basal Grits   } Grit

      CL – Main Limestone   } Carboniferous
      LLSh – Lower Lst. Shales } Limestone

      GG – Grey Grits
      B – Brownstones etc } 'Old Red
      RM – Red Marls   } Sandstone
      TB – Trichrug Beds (In the Silurian)

parison relatively soft and give rise to pronounced hollows and valleys.

The geological structure of the Black Mountains region is relatively straightforward (Figure 30), the strata in the main all dipping to the south. This is the result of the two major regional structures, the Towy Anticline and the South Wales Coalfield Syncline, on the northern and southern limits, respectively, of the route. This, then, with the fairly steep dips of the strata, accounts for the rapidity with which one traverses across four main geological systems. There is one important complication to the geological structure. Running through the area in a WSW-ENE direction is an important, but narrow, zone of folding and thrusting known as the Careg Cennen Disturbance. This belt of buckling and shattering of the rocks probably extends to Shropshire, and, as has already been seen, it exists in Pembrokeshire (the Johnston Thrust zone). Careg Cennen Castle is perched on a limestone bluff caught between two faults of this disturbance

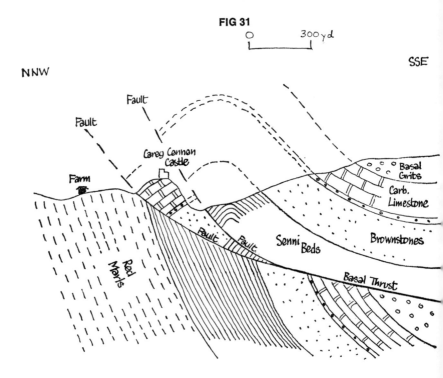

FIG 31

O   300 yd

FIG 32

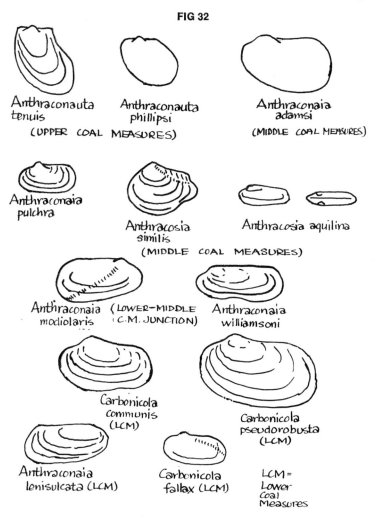

Anthraconauta tenuis

Anthraconauta phillipsi

(UPPER COAL MEASURES)

Anthraconaia adamsi

(MIDDLE COAL MEASURES)

Anthraconaia pulchra

Anthracosia similis

Anthracosia aquilina

(MIDDLE COAL MEASURES)

Anthraconaia modiolaris   (LOWER-MIDDLE C.M. JUNCTION)

Anthraconaia williamsoni

Carbonicola communis (LCM)

Carbonicola pseudorobusta (LCM)

Anthraconaia lenisulcata (LCM)

Carbonicola fallax (LCM)

LCM = Lower Coal Measures

(Figure 31). The Towy anticline, in the north of the Black Mountains traverse, is one of the main upfolded structures of Wales and is responsible for bringing an extensive belt of Ordovician rocks to the surface in the Vale of Towy. Many subsidiary faults cross the Black Mountains area in a NNW-SSE direction, and have complicated mining operations in the rich anthracite coal belt of Cross Hands and Ammanford. The surface effects of one of these main cross-faults will be described at a point near the summits of the Black Mountains (localities **4** to **7**).

Beginning the route at Swansea, one eventually comes to the town of Pontardawe and here a turn is taken to the left, along the valley of the Upper Clydach (or Gors). Nearing Cwmgors (Figure 29) the large modern colliery of Abernant is seen. On the hillside near here an important borehole was sunk in the 1920s, a borehole which proved the presence of shales of marine origin within the middle portion of the Coal Measures. The marine layer has become known, internationally, as the Cwmgors Marine Band and its equivalents have been proved in many other British coalfields. Nearing the village of Gwaun-Cae-Gurwen ('Gurwen's meadow'), a large coal tip looms up nearby (on the eastern limits of the village). A search on this tip will yield Coal Measure lamellibranchs ('mussels'), probably from the roof of the 'Brass Vein', one of the important anthracite coals of this area. They include *Anthraconaia modiolaris* (Figure 32) and smaller rotund forms.

Proceeding through Brynamman and taking the road to Llangadog (A4069), one begins the long ascent of the Black Mountains. Approaching the locality marked **1** in Figure 29, it is worth pausing and looking back to see the marked hollow of the softer lower half of the Coal Measures, with the harder Pennant sandstones forming the plateaux to the south. Within the main hollow a subsidiary ridge (between Brynammmam and Gwaun-Cae-Gurwen) is due to a thin but tough quartzite band called 'the Cockshot Rock'. The lower coals of this region and others further west in Carmarthenshire are high-grade anthracite coals. These have a very high carbon (and a low volatile) content, the gases having been largely driven off. Heat is believed to be the chief mechanism in the formation of anthracite, but the source of that heat is the subject of several conflicting theories. One theory says there is an igneous mass beneath this part of the coalfield. The presence of mineral veins (copper, lead and zinc) near Carmarthen and Kidwelly is cited as support for this theory. Another theory attributes the extra heat to the 'load' of higher Coal Measure strata above the rocks now seen. It is suggested that there was once another 8,000 feet of Coal Measures above those of this north-western portion of the coalfield. The third theory attributes the heat to friction caused by thrusting along the Careg Cennen Disturbance (Figure 31). This frictional heat would affect those coals of the coalfield nearest to the thrust plane, that is the coals of the north-western portion of the main coalfield. It is significant that the highly folded and thrust coals of the Pembrokeshire coalfield are also anthracites. On the other hand, the occurrence of pebbles of anthracite in Pennant sandstones presents problems, unless coals of some southern area had already been deformed before the deposition of the Pennant in South Wales. Again, in Somerset (near the town of Vobster) thrusting has pushed Carboniferous Limestone *over* the Coal Measures. These coals should therefore be anthracites, but, surprisingly, they are bituminous coals with a much higher volatile and gas content. No wonder that the 'Anthracite Problem' is still the '64,000 dollar question' in South Wales.

The bend in the Black Mountains road at Craig Derlwyn (locality 1) is alongside quarries in the 'Farewell Rock', the topmost unit of the Millstone Grit. The formation is so called because it lies below the lowest workable coals. Actually, the original meaning meant 'farewell to ironstone' because seams of clay-ironstone were worked in the South Wales Coalfield long before coal, with timber being used as the fuel in the smelting of iron. The Farewell Rock is formed of alternations of tough, wedge-bedded, siliceous sandstones with flaggy and shaley bands. Pebbles of clay-ironstone are sometimes seen in the sandstones. The formation was formed in a fluviatile environment, rivers pushing out into deltas fringing a southern sea somewhere over the western part of the Bristol Channel and Devon. The base of the Farewell Rock is seen around the bend to rest on shales, the topmost beds of the Middle Shales. A better exposure of these shales occurs some distance to the north (1a in Figure 29) but there are no fossil horizons exposed in this section. The grassy moorland of the shales gives way northwards to craggy country with bare exposures of light grey rock. This rock is very siliceous and is called 'quartzite'. This group of siliceous beds makes up the Basal Grits and their resistance is responsible for the Black Mountains. The high percentage of silica in some of the beds makes them suitable for the manufacture of silica bricks for lining steel furnaces. Some small quarries can be seen at locality 2. With almost 98 per cent silica in the rock, one would think that the other 2 per cent would be of no importance, but this is quite wrong. The alumina content must be less than 0.4 per cent in order to make a refractory brick that can stand up to modern furnace conditions. This is why the Dinas Mine near Glynneath closed down (see Chapter 8). Before leaving these almost white quartzites it is worth reflecting: why were these mountains called Black? The answer lies in the extensive peat covers, especially just east of the road.

The Basal Grits are underlain by the Carboniferous Limestone, first seen in a low quarry just east of the road at its highest point. This limestone has been much quarried for lime burning and as a flux in smelting iron. The narrow quarry road leading off to the east brings one to larger quarries (locality **3**). In these beds can be found brachiopods such as *Productus* and *Seminula* and the gastropod *Euomphalus* (Figure 48). In some of the burned, friable blocks of limestone, the Seminulas practically fall out of the rock. North of the quarries is a marked scarp and then a flat belt of very wet ground. This marsh is the outcrop of the 'Lower Limestone Shales', the basal unit of the Carboniferous Limestone. The northern edge of the marsh is a line of flat crags. A close examination of them (at the northern limit of the National Trust car park) shows them to be of a very coarse quartz-conglomerate with many pebbles of vein quartz and some of red jasper. These are the 'Grey Grit', the uppermost unit of the Old Red Sandstone (see Figure 30). They are not very thick but they make an easily recognised feature on the surface (locality **4**). They cross the road but then die out completely in about a hundred yards to the west. In fact still further to the west along the same line there are limestone quarries (locality **5**). The lowest of these is again in the dark *Seminula*-bearing beds. The top of this quarry is a very honeycombed rock with stringers of sand occurring within a limestone. This is the 'Honeycomb Sandstone' and is overlain by a very light grey 'oolite'—a limestone made up of tiny concentric balls of lime.

Why does the line of Grey Grits end between the road and these quarries? The answer is shown in Figure 34. A major fault —the Cwmllynfell Fault—crosses the region in almost a N-S direction and faults the conglomerate on the east against the limestone on the west. Standing on the conglomerate at the point of its disappearance, a grass filled gully becomes apparent, leading down the slope and beyond the road below. This is the line of the fault. The line of conglomerate crags on the *western* side

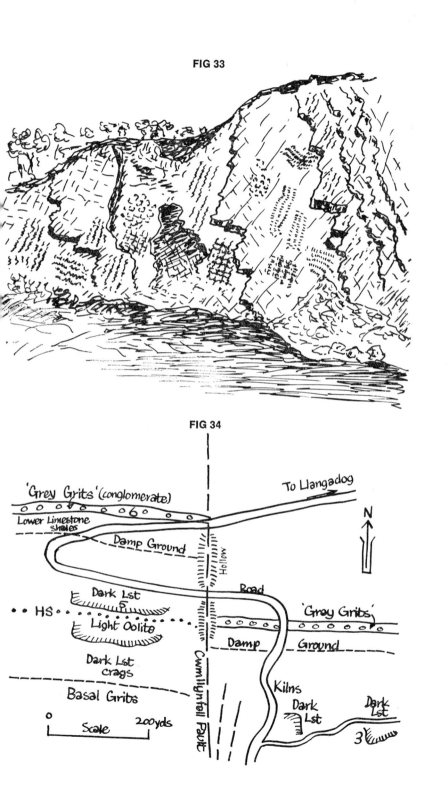

FIG 33

FIG 34

'Grey Grits' (conglomerate)

To Llangadog

Lower Limestone
shales

N

Damp Ground

Hollow

Road

Dark Lst

'Grey Grits'

•• HS ••

Light Oolite

Damp Ground

Dark Lst
crags

Kilns

Basal Grits

Dark Lst

Dark Lst

0

Scale

200yds

Cwmllynfell Fault

of the fault can be seen way down below, to the north of the hairpin bend in the road ('the Cuckoo Bend'). Even the damp ground of the Lower Limestone Shales is there—in fact a spring emerging at the top of the shales has deposited a cloak of calcareous 'tufa' near the lower road. The displacement or 'throw' of the fault must be almost 200 feet, the western side having dropped relative to the eastern block. The fault is an important one and becomes known as the Rhydding Fault between Pontardawe and Neath. It is in fact the easternmost member of the displaced trough in Figure 56 (see Chapter 7).

The descent of the Black Mountains northwards now begins and at locality 7 the thick Brownstones division of the Old Red Sandstone can be seen. The dull red sandstones, shales and marls give rise to many cascades in the nearby streams. The rocks are, however, barren of fossils. These harder red rocks give way northwards to the Red Marls, forming lower, but often wooded, ground. The base of the Red Marls ('the Tilestones') can be seen behind the inn at the village of Pont ar Llechau, whilst the overlying Red Marls can be seen in the banks of the River Sawdde, above the bridge. The village is named 'the Bridge on the Slates', the thin-bedded, very micaceous Tilestones having been quarried in the past for roofing houses and farm buildings. They are fossiliferous, as can be seen in the old quarry behind the inn (locality 8a). Gastropods include *Holopella*, whilst the main lamellibranchs are *Grammysia* and *Modiolopsis*. One interesting feature of the Tilestones (now called the Long Quarry Beds, after a long excavation almost a mile long, north of Careg Cennen Castle) is the very micaceous character of these beds. Recent work has shown that these sediments were transported from the south and that an area of ancient altered rocks (mica-schists) was probably being eroded to supply the material. If this is so, then it could mean that the Old Red Sandstone (and its Tilestone base) rests on Precambrian metamorphic rocks over what is now the coalfield.

Walking over the bridge and along the narrow road to the left (down-river) brings one to a large quarry (locality 8b) in very steeply tilted strata (Figure 33). The rocks are of Upper Silurian age and comprise calcareous limestones, grey shales and mudstones, together with tough pebbly grits and conglomerates, the latter often red or purple and particularly well seen in the southern portion of the inner face. Fossils can be found by looking carefully at rock fragments on the floor and scree, more especially in the shale and calcareous rocks. The latter often weather to a chocolate colour on the outside and can be rotten, when the fossils can be easily seen. They include *Camarotoechia* and *Chonetes* (Figure 35), crinoid ossicles, *Loxonema, Grammysia Platychisma* and *Modiolopsis*. Trilobites have been found here,

FIG 35

a. Trinucleus   b. Didymograptus murchisoni
c. Climacograptus   d. Atrypa   e. Ogygia
f. D.bifidus   g. Orthis   h. Halysites
i. Nicolella   j. Camaratoechia   k. Leptaena
l. Calymene   m. Chonetes   n. Monograptus nilssoni

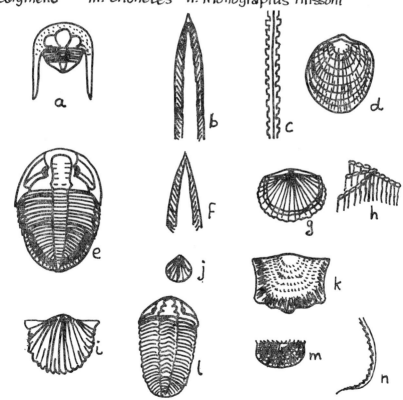

FIG 36

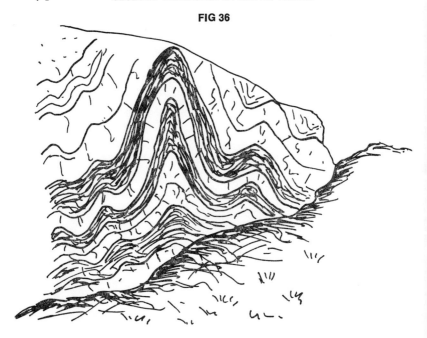

including *Calymene* (Figure 35). The outstanding feature of the quarry, however, is undoubtedly the well preserved ripple-marks on the steeply dipping bedding planes. Some of these are single parallel lines of ripples, others show cross-ripples. The ripples were of course made when the sediments were soft and wet and when the forming sediments were in a flat condition under water. These ripple-marked strata, now tilted up on end, are a good example of the power of crustal forces and demonstrate vividly the intense buckling of the Silurian and Ordovician strata along the Towy Anticline. The ripple-marks, together with the coarse pebbly conglomerates, indicate the shallow-water character of these 'Trichrug Beds'. They probably represent a thick delta pushing out into the Silurian Sea from a land area nearby. The striking ridge known as the Trichrug is caused by the resistance to erosion of the tough, pebbly grits and conglomerates, quarried not so long ago for roadstone.

Just over a mile north of Pont ar Llechau, a road to the left leads up a hill and a disused shallow quarry on the left comes into view. This quarry (locality **9**) exposes a light coloured siliceous grit—the Ffairfach Grit—in the middle of the Ordovician System. South of the quarry some indifferently exposed crags are seen. These are of volcanic rocks—rhyolitic ashes and tuffs, interbedded with shales. The ashes are not very thick and it is likely that the site of the volcanic extrusion was some considerable distance away.

Just south of the village of Llangadog another road leads to the left over a concrete bridge. This road leads to the village of Bethlehem. A small, rather overgrown quarry on the south side of the road (close to the house named Talar Wen) (Grid Ref SN 701266) exposes brown-weathering, calcareous flags, containing many trilobite fragments. These rocks are the Llandeilo Flags and occur just above the Ffairfach Grit, mentioned above. The two main trilobite genera found in this quarry are *Ogygia* (now called *Ogygiocarella*) and *Trinucleus* (Figure 35). Very fine specimens can be found showing the ornamented head borders of *Trinucleus*. The same rocks can be seen in a small quarry at the southern end of the railway cutting at Ffairfach (locality 11). Permission should be obtained before entry to this exposure, which yields, in addition to the trilobites, very large specimens of *Lingula*.

The return route to Swansea is along the A476 to Cross Hands and the A48 through Pontardulais. Two and a half miles south-south-west of Llandeilo the road climbs up through wooded country, and new road cuttings on the western side of the road expose the Upper Silurian and the overlying Devonian rocks. The dips in the strata are vertical and become even overturned by 20 to 30 degrees in the Old Red Sandstone. A gradual replacement of the brown-weathering strata by red beds is seen to occur in the section and true Red Marls are seen towards the southern end of the section. Coarse grits and conglomerates occur just below the base of the Devonian. The Silurian beds are fossiliferous and yield the brachiopods *Atrypa*, *Gypidula* and *Leptaena* (Figure 35) and the trilobite *Dalmanites*.

Continuing southwards around a wide bend (locality 14), views to the east begin to open up and in the distance the castle of Careg Cennen is seen on a prominent hill in the valley of the River Cennen (locality 12 in Figure 29). The castle can be visited by taking the road to Trapp from Ffairfach. Cars may be parked at the farm situated just north of the castle. The castle was the centre of rule of the Princes of Deheubarth in the twelfth and thirteenth centuries. Rhys Fychan, one of the princes, won back the castle in 1248 after his mother (Matilda de Breos of Gower) had betrayed it to the English out of pique for her son. Nine years later Rhys was thrown out of Careg Cennen by his uncle Meredudd. The English took the castle later that century and rebuilt it, but it seems to have been abandoned since the later fifteenth century. The castle overlooks a very steep rocky cliff above the River Cennen. The castle hill is of dark Carboniferous Limestone, but this limestone is completely surrounded by Old Red Sandstone, the limestone boundaries being faulted. The probable structure of the Careg Cennen hill-mass is shown in Figure 31. A major low-angled thrust-plane steepens upwards and splays into two branches, one on each side of the limestone mass. The possible influence of this major line of thrust on the coals of South Wales has already been discussed (p 68). Climbing the hill from the farm to the castle, one sees vertical Red Marls in the path. Just through the swing-gate is a major E-W hollow marking the position of the main thrust. The base of the castle's foundation exposes dark limestone with *Seminula* but the rocks are badly shattered. Eastwards, the grassy hollow of the fault becomes a deep ravine. Also badly shattered and disturbed are the Old Red Sandstone beds in the River Cennen, due to the presence of the southern (splay) member of the Careg Cennen fault system.

Continuing southwards along the A476, the village of Carmel is reached. The short slope north of that village is formed by the Brownstones of the Old Red Sandstone, but it will be appreciated that this Brownstone scarp is nothing in comparison with that of the Black Mountains and very much less than that on the Breconshire Fans and Beacons further east (Chapter 7 and Figure 54). The difference is due to the gradual cutting out westwards of much of the tough Brownstones succession by important unconformities which occur both at the base of the highest Devonian and at the base of the Carboniferous Limestone. This is really why there is no 2,000ft OD scarp to cross on the journey from Swansea to Carmarthen, in marked contrast to the more difficult journey from Merthyr Tydfil to Brecon.

South of the village of Carmel the junction of the Limestone and Millstone Grit occurs at a cross-roads, and the basal beds of the Grit are excellently exposed at a quarry 200 yards west of this cross-roads (locality **15**). This Allt y Garn Quarry shows a formation known as the 'Plastic Clays' which occurs only in this region at the base of the Millstone Grit. The rocks are alternations of fine grit with 'chert' (a form of silica) and very weathered chert in the form of greasy clay which really is plastic. The clays are dug for moulding and for repairing cracks in hearths, etc. Of particular interest in the quarry, at its western end, is the striking anticlinal fold which soars up to the top of the quarry (Figure 36). The origin of this structure is not clear. It could reflect a swelling of the wettened clays, it could be due to differential loading of these lubricated, incompetent beds, or it could be associated with faulting which is known to traverse this gap at Carmel.

The overlying quartzites of the Basal Grits are seen west of the road at locality **16**, with striking bedding planes of quartzites alternating with grassy hollows, marking shale intercalations. To the south and south-east is once more the important hollow of the Lower and Middle Coal Measures, with open-cast workings near the village of Penygroes. From Cross Hands to Swansea the route lies mainly over Pennant Measures, the alternating sandstone-shale character of which is responsible for the terrace effect of the hills overlooking Pontardulais.

Chapter 5

Roman Gold and Ancient Shorelines

The hills north of the Towy Valley appear to be carved out of two extensive high-level surfaces, one at 1,200-1,400 feet OD, the higher at 1,400-1,600 feet OD. The area is drained by two river systems, the Upper Towy (or Tywi) and the Cothi (which enters the main Towy system between Llandeilo and Carmarthen). The Cothi has many headwaters, including the Melinodwr, Morlais and the Twrch. The headwater region of the Upper Towy is a wild, rugged, dissected upland, noted in past days as the hiding place of a famous Welsh Robin Hood—'Twm-Shon-Catti' (his lair was in the crags just north-east of the confluence of the Upper Towy and the Doethie) but noted now for its new rock-filled dam and reservoir at Brianne (Grid Ref 798494). To the east lies the Irfon Valley, this river coming down swiftly from the hills to the ancient spa of Llanwrtyd Wells and flowing more gently eastwards to join the Wye at Builth Wells. The road from Llandovery to Llanwrtyd rises sharply in hairpin bends over the Sugar Loaf. Nearby, the much loved mid-Wales railway line to Shrewsbury buries itself in a tunnel through the mountain.

Geologically, this great hill mass north of the Towy lies on the north-western flank of the great NE-SW aligned Towy anticline. Along the eroded core of this great complex upfold, Ordovician strata are exposed. On the north-west flanks of the fold the lower Silurian (Llandovery) strata come on, but there is still a good deal of infolding, all these minor buckles trending NE-SW. Faults and great crushed belts trend the same way. The folded zones and crush belts are riddled with veins of vein quartz, and in some localities there occur ores of lead, zinc and copper pyrite, as for example at Rhandirmwyn, eight miles north of Llandovery. These minerals—the silvery galena (lead), the resinous coloured zinc-blende and the rainbow-sheened chalcopyrite—

can all be collected on the old tips half a mile north of Rhan-dirmwyn.

The most famous mineral veins of this district are undoubtedly, however, those of Ogofau ('caves') (or Dolaucothi). These old levels and mines are situated one mile east of the village of Pumpsaint ('Five Saints') and close to the left bank of the Cothi (see Figure 37). The old tunnels are there, only be careful how you proceed inside. Outside are the waste tips and here one can again collect. These Ogofau miners searched particularly for gold. The wedding rings of British royalty have on occasions been of Pumpsaint gold. The old workings cover an area of some five

FIG 37

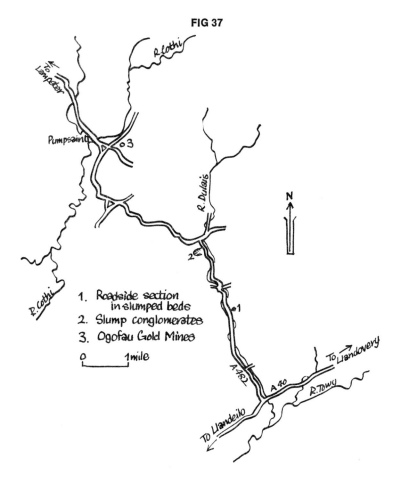

1. Roadside section in slumped beds
2. Slump conglomerates
3. Ogofau Gold Mines

0   1 mile

square miles. The Romans first worked here but whether they paid their labourers or whether they treated them as slaves is not known. The refining methods of the Romans certainly maintained a high, consistent purity of gold. Such purity today demands very complicated techniques. The Elizabethans worked here, as did several companies until fairly recently. The last venture was abandoned in 1939.

The gold-bearing quartz veins (deposited by hot igneous vapours penetrating upwards from below) occur in a 400 foot thick crush belt of intense crumpling at about the junction of the Ordovician and Silurian systems. Around the folds, aligned NE-SW, run the quartz veins. Across the folds trend N-S faults, carrying small quantities of galena and zinc-blende. Unfortunately, this fairly clear picture of the structure and mineralisation has been made more complex by an invasion of later quartz-veining, this time not bearing gold. Two fairly persistent gold-bearing lodes occur, however, in the upper half of the 400 foot crumpled belt—the Roman Lode and the Pyritic Lode. It is thought likely that further lodes may occur in the lower parts of the crush belt and on other belts nearby. There are, then, possibilities for this part of mid-Wales. The motto should therefore be: 'When in Pumpsaint, do as the Romans do.'

To learn something about the nature of these Upper Ordovician and lowest Silurian strata on this north-west flank of the Towy Anticline, visit a now disused quarry (locality 2 in Figure 37) just left of the A482 road (from Llanwrda to Pumpsaint) at a point four miles north-north-west of Llanwrda (Grid Ref 694367). This long E-W gash in the hillside exposes northward-dipping conglomerates in the uppermost Ordovician. The coarse conglomerates are interbedded with dark shales and mudstones. The conglomerates have a muddy matrix and are examples of 'slump conglomerates'. These coarse pebbly beds were initially deposited on a shallow continental shelf but sea-bed shocks subsequently shook the pebbly material and sent it sliding at speed down the steeper continental slopes, perhaps even being directed along the bottoms of sloping submarine canyons, such as those on the Newfoundland Banks today. This coarse material eventually settled into finer muds in much deeper water. Graptolites have been found in the interbedded shales of this quarry. One must not rigidly maintain, therefore, that coarse conglomerates always mean a shallow-water environment. The contortions in the strata caused by the down-slope sliding or 'slumping' can be seen in an exposure on the roadside (east side) two miles further south (Grid Ref 703346). The direction of slumping can be shown to be towards the north-west both in this exposure (Figure 37, locality 1) and in the quarry mentioned above (locality 2).

The Ordovician rocks of this region are lacking in volcanics, but a local occurrence of volcanics, tuffs and agglomerates occurs about two miles north of Llanwrtyd Wells (Garn Dwad and adjacent hills).

BUILTH WELLS

North-east of Llanwrtyd Wells the structure of the Towy Upfold becomes more complex and a parallel anticlinal structure makes its appearance in the high rugged ground immediately north-north-west of the town of Builth Wells. This NNE-striking ridge is known as Carneddau and is bounded to the west by the A483 road to Llandrindod and on the east by the A481 to New Radnor (Figure 38). The summit of the ridge reaches to over 1,400 feet OD. This high ground is formed of the thick Builth Volcanic Series, erupted during Llanvirn times in the Ordovician period. The volcanics total nearly 2,000 feet in thickness and comprise fine ashes, pebbly and bouldery ashes, agglomerates, and lavas of spilite and keratophyre. Both of these lava types are rich in soda, but whereas the keratophyres are of 'acid' type (rich in silica) the spilites are less rich in silica and therefore contain little quartz. Besides the extrusive volcanic rocks there are also intrusions, mostly of dolerite. These Llanvirn volcanics are overlain by softer black shales which give rise to the much more subdued and lower ground west of the A483. Here, then, is a good example of the difference in topographical outline and effect of resistant igneous rocks versus soft shales. The River Wye, too, seems to have done its best to skirt around the southern end of the ridge of hard volcanics, at Builth Wells.

The various volcanic horizons on the Carneddau dip steadily west at about 30 degrees so that their surface outcrops trend from north to south. Many faults break the continuity of these ridges, however, and most of these fractures trend E-W. Important examples are the Carneddau Fault-belt and the Wern-tô

FIG 38

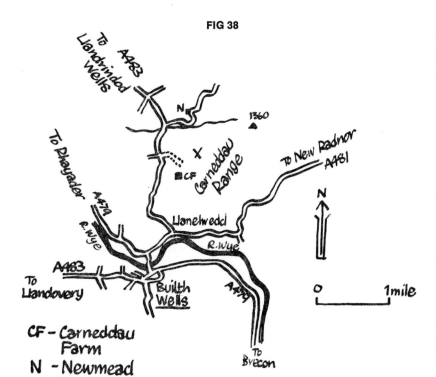

To A483
To Llandrindod Wells

To Rhayader

N

1360 ▲

Carneddau Range

X
CF

A479

R. Wye

To New Radnor
A481

N

Llanelwedd

R. Wye

A483

To Llandovery

Builth Wells

A479

To Brecon

0   1mile

CF – Carneddau Farm
N – Newmead

FIG 39

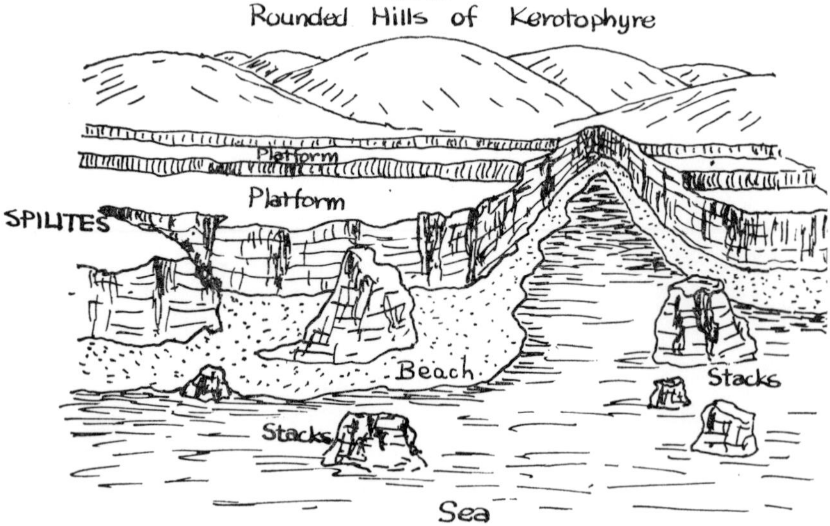

Rounded Hills of Kerotophyre

Platform

Platform

SPILITES

Beach

Stacks

Stacks

Stacks

Sea

(Looking Westward)

Fault (Figure 40a). In places the former fracture zone has a northerly downthrow amounting to over 1,000 feet. There are several working and disused quarries on the Carneddau. The hard volcanics and interbedded grits are excellent for road surfacing.

This Carneddau area is of interest not only because of its great variety of igneous rocks but also because it is possible to reconstruct the geography of this region in mid-Ordovician times. This reconstruction is largely the work of two eminent Welsh geologists—Sir William Pugh and the late Professor O. T. Jones. They mapped this region in very great detail and published their results in the journal of the London Geological Society in 1949. The account includes three detailed maps (coloured) and should be read by anyone who wishes to visit this area.

Their work has revealed the preservation over the Carneddau of an ancient coastline of mid-Ordovician age. About 470 million years ago there was a coastline here and its detail can still be made out. At one particular phase in its development that coast looked like the illustration in Figure 39 (this is a drawing from the reconstruction in the account by Jones and Pugh, 1949, p 79).

The geological story of the region is as follows: after the deposition of the lowest Llanvirn sediments (muds), great and prolonged vulcanicity broke out at Builth. At first the activity was of an explosive type, with ash and bomb-like agglomerates being laid down. Then came lava extrusion, probably over the sea bottom and the spilites accumulated, interrupted by further bomb phases. The last volcanics to accumulate were the acid keratophyres. These may even have been 'intruded' into the near-surface sheets.

There then followed emergence through uplift (but with no apparent tilting of the various layers of volcanics) and erosion of the land area (probably a large island or series of islands in the mid-Ordovician sea) commenced. The erosion of the spilites and ashes produced a 'trap' topography of terraces and scarps, closely resembling that in the Deccan of India or in the Azores today. The different reaction of the more silicified keratophyres produced rounded knoll-like hills on those volcanics. And so a topography like that of Figure 39 was produced with the sea still at hand. Then the sea level began to rise (or the island began to sink) and coastal sands spread in over more and more of the drowning land area. As the sea level rose, so each terrace became covered with water and sand covered each flat surface. These sands would become coarser and more bouldery when traced towards the bottoms of each scarp. Screes would fall from these cliff edges into the nearby water. Ultimately, even the knolls of the keratophyres were submerged beneath the sea and covered by sands and boulders.

These sandy and pebbly deposits, rising steadily up the flanks of the terraced island, are known as the Newmead Series. The sands are rich in felspar, but there is a difference between the sands deposited over the ashes and spilites and those over the keratophyres, the latter being more pyritic. Boulder beds can occur in both types of sand, the boulders varying from one to four feet in size. An exposure which is particularly interesting is a quarry 100 yards east of the farm called Tan-y-graig (just over half a mile north of the Llanelwedd junction of A481 and A483). The upper part of this exposure is in grey felspar sands, containing near the top a rusty band with brachiopods and fossil sponges. The sands have been stripped down to a boulder-bed surface. A single detached boulder lies on the floor of the quarry. This boulder has a smooth flat back surface but the other side is fluted. This must be the side of the boulder which was subjected to wave attack when the boulder formed part of a cliff on this Builth coastline in the Ordovician period. The top of the boulder bed extends along the slope to the south for over three hundred yards.

Figure 40a shows the relation of the sandy deposits and boulder beds to underlying volcanics in the area between the Carneddau and Wern-to Fault belts. The section in Figure 40b is drawn to show the terraced contact of the Newmead sands and the volcanics as it appears today, that is after the westward tilting of the Ordovician strata. To see this section as it would have looked *before* that tilting, turn the page so that the junction of the spilites with the keratophyres is horizontal. The Newmead sands can then be seen to rise up the terraced spilites and bury the keratophyre hills. At **B** in Figure 40b the keratophyre emerges. At **C** the breccias represent the scree debris on the eastern side of a keratophyre hill. At **E**, **F**, **G** and **H** the sand cover has been eroded to expose the once sand-covered spilites, etc.

FIG 40

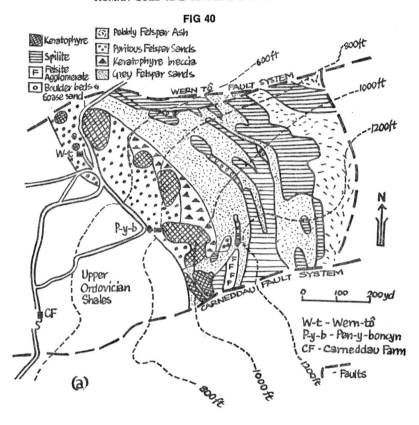

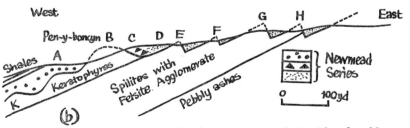

The large quarry immediately north-east of the Llanelwedd road junction of A483 and A481 shows a good range of the Builth Volcanic Series, from Lower Spilite at the eastern end through agglomerate, Upper Spilite, andesite and keratophyre as one proceeds westwards across the quarry. There is a thick dolerite intrusion within the Upper Spilite.

OLD RADNOR

The A481 from Builth to New Radnor climbs up over hilly country carved out of Silurian grits and shales. On meeting the A44 (Rhayader-Kington) road, it then descends down the valley of the Summergil Brook to the medieval town of New Radnor. Three miles east-south-east of that town there occurs another fascinating geological area, that of Old Radnor and Dolyhir. Here there are inliers of Precambrian rocks (igneous and sedimentary), appearing from beneath a cover of Silurian sediments (Figure 41). There are no Cambrian or Ordovician rocks present. Presumably they were once present but were uplifted and worn off prior to the deposition of the Silurian. It is therefore worth reflecting that the thick pile of Ordovician rocks at Builth may have been denuded completely here before Silurian times began. The Old Radnor area lies on a southerly continuation of the Church Stretton Disturbance, an area in which extensive exposures of Precambrian rocks occur, including the high Longmynd plateau and the striking igneous hills of Caer Caradoc and the Wrekin. The Church Stretton Disturbance probably continues through the Old Radnor region and on through Breconshire into Carmarthenshire.

At Old Radnor the Precambrian rocks have been upfaulted along two distinct belts of high ground (see Figures 41 and 42). The western belt is formed of Precambrian sedimentary rocks (the Longmyndian) whilst the eastern belt, comprising the striking hills of Stanner Rocks, Worsell Wood and Hanter Hill, is formed of igneous rocks, dolerites, gabbro and acid porphyries, probably of the same Precambrian age as the Uriconian of Shropshire. The two belts are very much fault-bounded (see Figure 41). They are both upfaulted 'horsts'. Between them lies a tract of lower ground in softer Wenlock shales and siltstone. Old Radnor Hill, formed of Longmyndian grits, flags and slates, reaches a height of 1,081 feet OD. These Precambrian ancient sediments are well exposed in the Gore Farm Quarry, half a mile east of Old Radnor Church. In one corner of this quarry is exposed one of the numerous faults in this area.

FIG 41

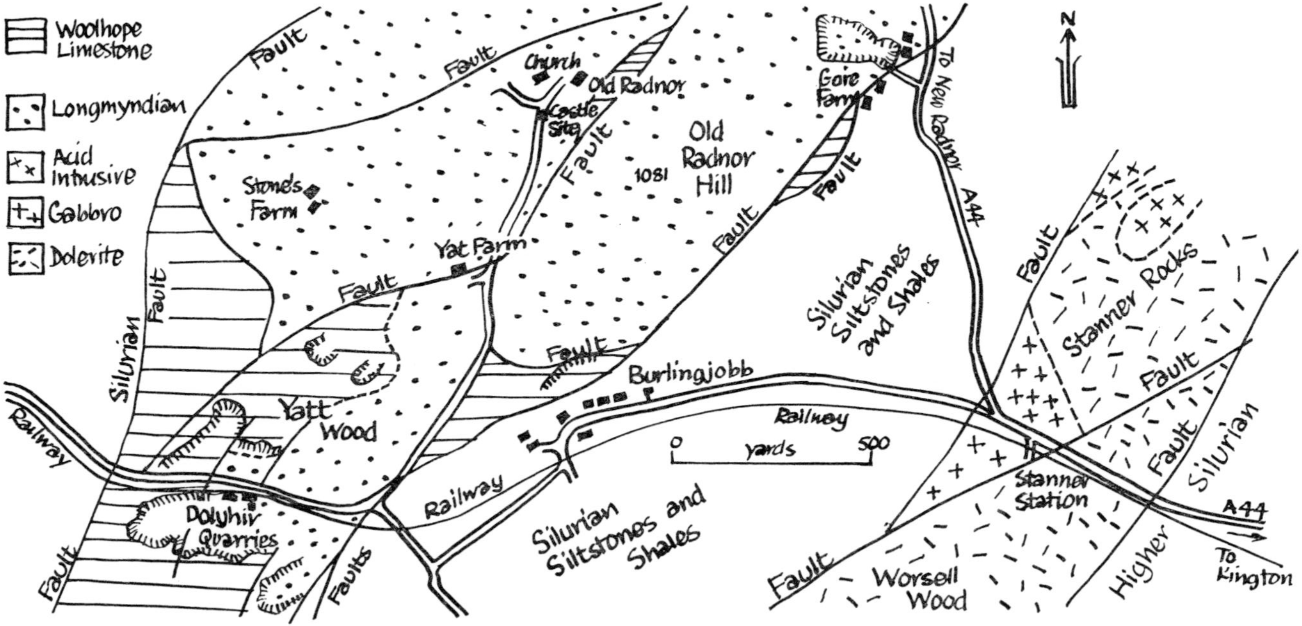

N

To Kington
A44
Higher Silurian
Fault
Fault
Fault
Stanner Rocks
Fault
Stanner Station
Worsell Wood

To New Radnor A44
Gore Farm
Fault
Silurian Siltstones and Shales
Old Radnor Hill
1081
Fault
Fault
Burtingjobb
Railway
Silurian Siltstones and Shales
Railway
0 yards 500

Old Radnor
Church
Castle
Swg
Fault
Fault
Yat Farm
Fault
Fault
Railway
Faults
Fault

Fault
Stonds Farm
Fault
Yatt Wood
Fault
Dolyhir
Gwernie
Silurian Fault
Railway

Woolhope Limestone
Longmyndian
Acid Intrusive
Gabbro
Dolerite

Typical of the Precambrian igneous rocks is the coarse mottled gabbro seen on the southern slopes of Stanner Rocks (opposite Stanner Station). The gabbro is intruded into a less crystalline dolerite and the contact of the two igneous intrusives can be made out on this slope. Lighter coloured intrusive rocks occur nearer to the summit of Stanner. Worsell Wood, on the south side of the main road, is entirely formed of dolerite. At one time it was believed that these igneous rocks were intruded into a Silurian cover, but it is now thought that the igneous masses are of Precambrian age, brought up by faulting along the Church Stretton belt. This view is supported by the remarkable similarity between the geological pattern of the Church Stretton and Old Radnor areas, with Longmyndian, Silurian and Uriconian out-cropping, in that *eastward* order, in both districts.

The contact of the Precambrian floor with the overlying Silurian is best seen in the large Yat Hill Quarry (Grid Ref 242583) north of the railway at Dolyhir (Figure 43). The Silurian succession begins with the Woolhope Limestone, a massive light grey limestone of almost pure calcium carbonate. In other areas of Wales and the Welsh borderland rocks of this age are underlain by earlier Silurian formations, but these are missing here in the Old Radnor district, presumably because they were never deposited, the region being then one of uplift and erosion. The Woolhope Limestone in Yat Hill quarry dips at relatively gentle angles, but the underlying Longmyndian is highly folded in places. It is rather surprising, in view of the great gap in time between the Precambrian and the overlying Silurian limestones, that there is not a thick development of basal conglomerate or sandstone at the base of the Woolhope Limestone. It must mean that after long denudation to a virtually flat, peneplained surface, the region sank slowly beneath a clear warm sea—a shelf sea that extended into the Welsh borderland. To the west were deeper waters with submarine slopes and trenches, down which plunged coarser sands, pebbles and muds (the Silurian of the Rhayader and Cardiganshire areas). Actually, in places there is a thin conglomeratic development at the base of the Woolhope Limestone in the Old Radnor area. It can be seen in an old quarry 300 yards south-west of Yat Farm. The basal conglomerate is about 9 feet thick and consists of pebbles of quartz and grey-wacke, up to 2 inches across, set in a matrix of dark grey lime-

FIG 42

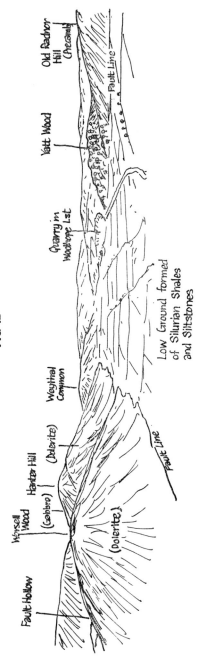

(LOOKING SOUTHWESTWARDS)

FIG 43

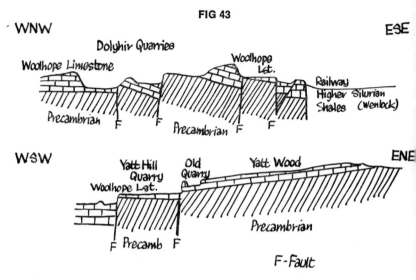

stone with a good deal of secondary calcite. The Precambrian floor cannot exactly be seen here, but it can be seen to outcrop in the wood nearby.

The Woolhope Limestone has been extensively quarried south of the railway at Dolyhir. The contact of the limestone formation and the Precambrian is here largely a thrust-plane inclined at 20 degrees. Above the thrust is a considerable thickness of the massive limestone. Below it, and seen especially in a lower quarry level, is the Precambrian—greywackes with conglomerate bands. The thrust-plane is itself faulted, with throws of up to 25 feet. Fossils are common in the Woolhope Limestone. They include corals such as *Favosites*, *Cistiphyllum*, *Halysites* and *Streptalasma*. Crinoid stems are common. Brachiopods include *Athyris*, *Atrypa*, *Leptaena*, *Pentamerus*, *Strophomena* and *Rhynconella*. The nautiloid cephalopod *Orthoceras* also occurs. Trilobites such as *Illaenus* and *Acidaspis* have been found here. In places, masses of the tabulate corals *Favosites* and *Halysites*, together with bryozoa, abound, so that the area was virtually a coral reef at that time.

According to local records, the limestone of this district has been worked for over 350 years. In earlier days it was in demand as a smelting flux. Nowadays it is used mainly for agricultural purposes.

THE RESERVOIR COUNTRY

The town of Rhayader is situated some eighteen miles north-west of the Old Radnor area. The River Wye flows in a narrow valley through Rhayader, on its way towards Builth Wells. The ground to the west and north of Rhayader is rugged and reaches heights of 1,400 to 1,800 feet OD. Two tributaries, in particular, of the Wye drain this rugged region—the Elan and the Claerwen. Their deep winding valleys make them ideal for reservoir construction. The two rivers are fed by innumerable fast-flowing tributaries and headwaters and there is an abundant rainfall throughout the year. The Silurian strata of the hillsides include thick, tough, coarse conglomerates which can be quarried, near to the dam sites, for the building of the rock dams.

There are some five main reservoirs—Caban-coch is the one nearest to Rhayader, Craig-yr-allt-goch and Claerwen are the two furthest away. There is a magnificent ride along the road that winds around the reservoirs, and one can continue along a wild mountain road that will eventually lead one to Devil's Bridge and Aberystwyth.

The reservoir area is carved out of Lower Silurian (Llandovery) strata. These are all of sedimentary origin and comprise shales, flags and mudstones interbedded with thick local developments of coarse pebbly conglomerates. One conglomerate formation— the Caban Conglomerate—is up to 600 feet thick in places. It is well developed on the ridge of Craig Cnwch (overlooking the dam of the lowest reservoir). The rock for this dam was quarried on the northern slopes above the dam. The conglomerates are well exposed even near the road. They are once again 'slump' conglomerates which were carried by turbidity currents into deeper water. Dr Gilbert Kelling has shown that the material was carried along a deep submarine canyon which ran north-westwards in this Rhayader area. This supports the view expressed above that the Old Radnor area lay in the quieter marginal shelf sea of Silurian times. That eastern area was probably a much peneplained lowland when the Caban pebbly masses were being eroded off higher ground somewhere to the south of Rhayader and being swirled by fast-flowing submarine currents down the troughs of these Silurian submarine canyons.

The structure of the reservoir region comprises broad folds trending NNE-SSW to NE-SW. One major anticline brings up the Upper Ordovician to the surface south-west of the Claerwen Valley. Some important faults occur. One runs N-S some 200 yards west of the Caban-coch dam, another occurs about one mile further west. Geologists and engineers have had to pay great attention to the location of these faults. It would be no good constructing a great rock dam on top of a fault belt. Faults which cross more central parts of reservoirs are all right because any water draining or upwelling along those fractures can be incorporated in the reservoir basin. Geology obviously plays an important part in the siting of engineering constructions such as reservoir dams, roads, road cuttings, buildings, etc. Geology is a practical as well as a cultural subject and its contribution is rapidly being appreciated today.

CHAPTER 6

The Gower Peninsula

Gower is an east-west elongated peninsula, connected to the South Wales 'mainland' in the area between Swansea and Loughor. The peninsula measures sixteen miles (E-W) by eight miles (N-S). It has a much indented southern coastline with numerous attractive bays (Figure 44), the largest being those at Oxwich and Port Eynon. The western edge of the peninsula is dominated by the broad sweep of Rhosili Bay with its two projecting arms—Burry Holms and Worms Head. The latter really looks like a winding serpent when viewed from the air. Swansea Bay dominates the east side of the Gower Peninsula and the projecting headland isles of Mumbles dominate the seaward view from the hills around the city of Swansea.

There is a marked difference in the character of the northern and southern coasts of Gower. The south coast is rocky with many steep cliffs, whereas the northern side is dominated by the broad sand and mud flats of the Burry estuary, through which at low tides meander the rivers Loughor, Lliw and Llan. The projecting spit of Whiteford Point marks the real edge of Carmarthen Bay on this north-western side of the peninsula.

THE SURFACE OF GOWER

The topographical surface of Gower can be generally subdivided into three main levels—200, 400 and 600 feet above sea level. These levels represent the present heights of surfaces which were cut by earlier sea levels. They are therefore uplifted marine or wave-cut platforms. Few remnants of the highest platform now remain, this surface having been largely 'eaten into' during the formation of the lower surfaces. The summits of the highest Gower hills—Rhosili Down (632 feet), Cefn Bryn (609) and

FIG 44

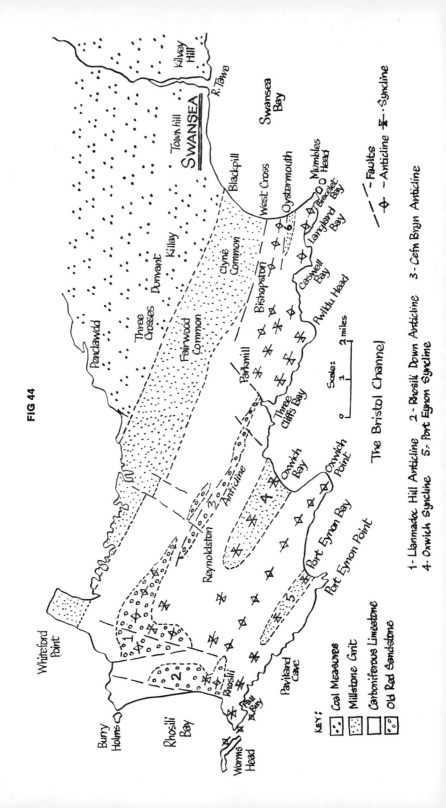

KEY:

Coal Measures

Millstone Grit

Carboniferous Limestone

Old Red Sandstone

Scale: 0 __ 1 __ 2 miles

The Bristol Channel

/ - Faults   ♦ - Anticline   ☲ - Syncline

1 - Llanmadoc Hill Anticline   2 - Rhosili Down Anticline   3 - Cefn Bryn Anticline

4 - Oxwich Syncline   5 - Port Eynon Syncline

Llanmadoc Hill (609)—represent these last modern relics of a
surface cut by the sea probably about 3 million years ago. The
two hill masses which dominate the north side of the city of
Swansea—Townhill and Kilvey Hill—are further vestiges of this
'600 foot Platform'. Remains of the 400 foot surface occur in
North Gower, in the vicinity of the village of Penclawdd (noted
for its cockles and laverbread) and Three Crosses, and again on
Clyne Common in eastern Gower.

The most extensive surface is however the 200 foot Platform,
which is particularly well preserved around the cliffs of Caswell,
Langland and Mumbles. The steeply dipping limestone beds on
the west side of Caswell Bay are bevelled by a remarkably flat
surface along the cliff top and a fine view of this uplifted marine-
cut surface is obtained as one descends into Caswell Bay from
its eastern (Newton) side. Other examples can be seen from the
cliff tops to the south of the village of Rhosili.

Some thorny problems are still connected with the origin of
these Gower, and other British, coastal platforms. Were they
the result of a static sea level and a rising land, or vice-versa?
Many people now think that it was the sea level that was
oscillating. The best answer, however, may well be a comprom-
ise. Again, where are the deposits of these marine platforms?
Maybe they were removed by the later ice sheets.

THE ROCKS OF GOWER

The simplicity of this surface pattern of Gower masks the
complexity of the geology and geological structure of the peninsula. The rocks of Gower are folded and much fractured and as
a result a variety of geological formation occurs. Rocks varying
in age from Devonian (360 million years old) to Upper Carboniferous (290 million) are exposed and there is even one important
patch of Triassic (200 million years old) in the Port Eynon area.
The rock succession seen in Gower is therefore:

Triassic breccias and marls	thin
(great break or unconformity)	
Upper Coal Measures	5,000 feet
Middle Coal Measures	2,000 feet
Lower Coal Measures	2,000 feet
Millstone Grit	2,000 feet

D

| Carboniferous Limestone | 4,000 feet |
| Old Red Sandstone (Devonian) | 1,000 feet seen |

The Devonian strata seen in Gower are coarse pebbly conglomerates overlying red sandstones and marls, probably all deposits formed under fluviatile or estuarine conditions some 360 million years ago and at a time when Britain lay close to the Equator and was experiencing a tropical (probably monsoonal) climate. The succeeding Carboniferous Limestone marks a great change in the picture over South Wales, the meandering rivers of Devonian times giving way to a shallow warm sea fringed by coral reefs and a fairly low-lying northern landmass.

The Millstone Grit represents another fairly sudden change in conditions, with rivers again invading the area with mud and sand. Gower probably lay near to the sea's edge and much of Gower's 'Millstone Grit' is a misnomer in that the succession is nearly all made up of muddy shale, which one can see in the banks of the Bishopston stream on Barland Common.

The broad belt of Millstone Grit across northern Gower from West Cross to Whiteford Point is succeeded northwards by the Coal Measures of the main South Wales Coalfield. The Lower and Middle Coal Measures are alternations of shales with occasional bands of sandstone. Seams of coal, each underlain by beds of fireclay, occur at intervals and were worked earlier this century along Clyne Valley in the east and near Three Crosses and Penclawdd further west. Forested swamps and mudflats with meandering sluggish rivers lay over South Wales at this time. The Upper Coal Measures, however, are dominated by the rusty-weathering Pennant sandstones, well seen near the new bridge over the River Neath, on the approach to Swansea from Port Talbot. These sandy, and occasionally pebbly, sediments were deposited (probably rapidly) by rivers flowing northwards from a mountainous area somewhere over south-west England. These Pennant sandstones are blue-grey when fresh but weather rusty (due to their iron content). They account for the hilly country immediately north of the city of Swansea and can be seen in quarries at Dunvant and Cockett.

At the close of Carboniferous times the flat sheets of Devonian and Carboniferous deposits were buckled, fractured and dislocated by the powerful Armorican earth movements. Some areas, such as Cefn Bryn, were upfolded, probably to appreciable

FIG 45

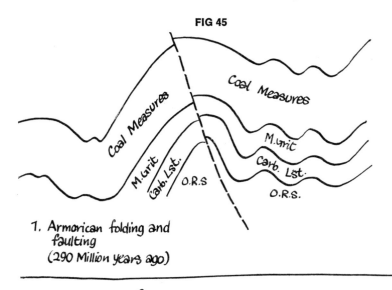

1. Armorican folding and
   faulting
   (290 Million years ago)

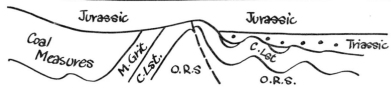

2. Erosion of folds and deposition of Mesozoic blanket

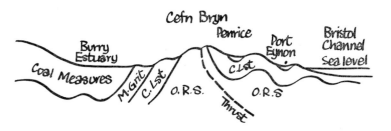

3. Tertiary and Quaternary erosion.
   Removal of the Mesozoic cover
   (except for some Triassic at Port Eynon)

heights. Others, such as Oxwich and Port Eynon, were down-folded into synclines. Erosion at later times—more especially in the Permian and Triassic periods and again during the Pliocene marine plantations—has ultimately bevelled the modern surface of Gower across these folded strata, producing the diverse geological pattern of today. Important events in Gower's past geological history are summed up in Figure 45, showing how today the Old Red Sandstone is exposed on Cefn Bryn whilst the much higher (depositionally) Millstone Grit occurs in Oxwich Bay. Over the worn-down folds of Gower were deposited Triassic, and probably later Mesozoic strata, only to be almost completely removed again during Tertiary and Pleistocene erosion.

MUMBLES HEAD TO CASWELL BAY

The cliffs from Oystermouth to Caswell Bay provide wonderful exposures of the upper half of the Carboniferous Limestone rock succession together with spectacular views of the folds and faults of this area. This coastal area is very accessible from Swansea with frequent bus services, especially in the summer. A coastal path from Limeslade Bay to Langland Bay and again from Langland to Caswell affords wonderful views of this attractive 'imestone coast. Four bays occur in this coastal stretch —Bracelet, Limeslade, Langland (with its eastern recess, Rother-slade) and Caswell. Each bay owes its existence to lines of weakness in the strata—faults or joints, weaknesses which have been exposed by the great erosive action of the sea.

The Limestone succession in this south-eastern corner of Gower is as follows:

'The Black Lias'	300 feet
Coarse Crinoidal Limestones with pseudobreccias	600 feet
'The Seminula Oolite'	800 feet
Coarse Crinoidal Limestones	400 feet
'The Modiola Phase'	20 feet
'The Caninia Oolite'	100 feet
'The Laminosa Dolomite'	100 feet seen

(The still lower parts of the Limestone succession can be seen further west in the Three Cliffs Bay and Rhosili—and will be described there.)

FIG 46

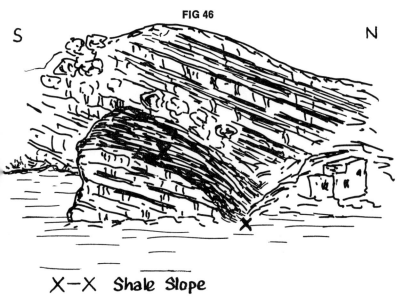

S   N

X—X   Shale Slope

FIG 47

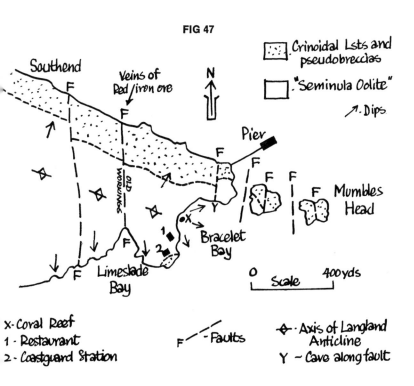

Crinoidal Lsts and pseudobreccias

"Seminula Oolite"

⌐ · Dips

Southend

Veins of Red /iron ore

N

OLD WORKINGS

Pier

F   F   Mumbles Head

Bracelet Bay

1   X

2

Limeslade Bay

0   Scale   400yds

X- Coral Reef
1 - Restaurant
2 - Coastguard Station

F — Faults

⊕ · Axis of Langland Anticline
Y ~ Cave along fault

'The Black Lias Quarry'   The northernmost exposure of the Limestone at Oystermouth is of 'the Black Lias'. This 'Clements Quarry' (Figure 46) lies just east of Oystermouth Castle and has a narrow entrance from the main road, just opposite the car park and the northern approach to the bus centre. These uppermost beds of the Carboniferous Limestone are alternations of dark muddy limestones with thin dark calcareous shales. The limestones weather with a white skin and 'ring' when knocked with a hammer. The 'ring' is like that of a thin china cup—hence the slang name of 'chinastone' for this rock type. These limestones have been used in the building of many of the old walls in Oystermouth, at one time a small fishing and oyster-catching village, whose steep streets climbing up the limestone cliffs can still be seen. The stone has also been used in the initial building (in AD 1100) of Oystermouth Castle and in its many phases of later rebuilding, consequent upon fierce Welsh uprisings.

The resemblance of the alternating limestone-shale sequence in this quarry to the Lower Jurassic 'Lias' accounts for the name 'Black Lias' and the same problems surround this Carboniferous rock formation as its Mesozoic 'cousin'. For example, were the limestone bands formed individually to be overlain by each shaley layer or was the deposit originally one thick mud with richer calcium carbonate bands ultimately 'drying out' to form the harder limestone layers? There is evidence in favour of both theories. For example, some limestone layers have cracks on the upper surfaces filled with mud from the overlying shale. Other limestone layers, however, have fossils which continue into the overlying shale.

The shale bands are quite fossiliferous, especially the shale band that forms the long, inclined, exposed bed in the quarry (**X** in Figure 46). As one walks up this shaley slope, over fossil shells, one is of course walking on an ancient sea bed of 320 million years ago, but a sea bed that was once flat and is now inclined at an angle of twenty degrees. Brachiopods are the most common fossils here, especially *Martinia* and *Spirifer* (Figure 48). One of the species of *Spirifer* is named after Oystermouth— *Spirifer oystermouthensis*, as also is the small cornet-shaped coral —*Zaphentis oystermouthensis*. Rare but important finds in Clements Quarry are of trilobites, one genus of which is *Griffithsides*.

One further point of interest about the quarry is the way in which the tilt or dip of the strata curves round within the quarry. This is due to the quarry being cut into the 'nose' or end of a pitching anticline. This anticlinal fold is separated from another which dominates the Mumbles Head-Langland-Caswell area by the Oystermouth Syncline, which is a downfold preserving softer Millstone Grit shales under Oystermouth (or Underhill) Park. If one visits that park, one can see the high surrounding rim of hard limestone with the great hollow eroded into the softer overlying, but downfolded, shales of the Millstone Grit. These shales are now only exposed (and even here rather poorly) on the hill that climbs up to the village of Newton.

Mumbles Head and Bracelet The cliffs at Southend (just beyond the moorings of the sailing boats) are of very steeply dipping limestones, more massive than those of the Black Lias Quarry. The steep dip along this northern side of the Mumbles promontory is in the northern limb of the Langland Anticline (Figure 47). Coarse crinoidal limestones (full of fragments of crinoids or 'sea lilies') are here interbedded with peculiar bands of reddish or greenish clay. These clays rest on very irregular surfaces of limestone. Where the clays have been removed the bare underlying limestone surfaces show many irregularities and hollows— admirable places for the location of flood-lighting equipment and illuminated gnomes during the summer illuminations. A search over the limestone surfaces along the road just beyond the Yacht Club is rewarded with finds of the large single coral *Palaeosmilia murchisoni* (named after one of the great geologists of the last century, Sir Roderick Murchison) together with the large gastropod *Euomphalus* (Figure 48). Veins of cream-coloured calcite, often reddened with iron oxide, are also common. Some of the oxide veins are sufficiently rich in 'haematite' to have been worked in the past. The best example is along 'the Cut'—the wide gulley in the cliffs beyond the Yacht Club. This vein was even worked by the Romans, the vein being along a fault which comes out on the other side of the headland to form the narrow inlet of Limeslade Bay. The iron oxide in, and reddening of, the limestones is due to downward percolation from a one-time red Triassic cover, now removed by erosion.

The road cutting leading to Mumbles Head reveals one of the clay bands mentioned above but also peculiar blotchy grey and

FIG 48

Palaeosmilia
(A single coral)

Martinia
(A brachiopod)

Seminula
(A brachiopod)

Chonetes
(A brachiopod)

Euomphalus
(A gastropod)

Spirifer
(A brachiopod)

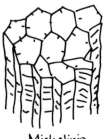

Michelinia
(A tabulate coral)

Camarotoechia
(A brachiopod)

Patella
(Recent gastropod)

fawn limestones with numerous dark blotches or fragments. This rock is called a 'pseudobreccia' and its origin is still a puzzle. There is also much calcite or 'spar' along this cutting.

Beyond the cutting (and the shop built like a cider apple) lies Bracelet Bay, the first (from the east) of Gower's bays (Figure 47). Standing on the lip of the long car park space here, one can see the cause of the bay's location. The limestone beds curve round in a great semicircle in the bay. This is the plunging nose of the Langland Anticline. When this fold was being bent by the Armorican earth forces, cracks occurred in the bending beds as they were being stretched over the top of the fold. The sea has detected these weak cracks and carved a bay into the coast here. Along some of the cracks (or 'joints') slip movements have occurred. One such fault occurs at the cave in the north corner of the bay. The calcite-lined walls of this fault are covered with grooves or 'slickensides' showing that the slip was along a horizontal direction. Another fracture separates the Lighthouse island from the middle island (itself bisected by yet a third fault).

Two other features of interest occur in Bracelet Bay. The rocks of the north-facing slope beneath the coastguard station are rich in fossil algae (calcareous seaweeds)—circular, ten-pence sized, concentric bodies—and the brachiopod *Composita* (or *Seminula*). The alternative name for this shell is used in the name of this limestone formation—the Seminula Oolite. The markedly oolitic texture in these oolites can be studied in the central portion of the bay where there is also a fossil coral reef, rich in a circular coral called *Chaetetes*. This coral reaches almost a foot in diameter and has been excavated by high tides into deep potholes.

Langland Bay This is perhaps the most majestic of Gower's bays, especially when viewed from the sharp road bend high on its eastern side. Langland's detailed outline is, however, complex, due to its being eroded by the sea into a zone of some five faults trending at right angles to the coast. The easternmost fracture is that responsible for Rotherslade Bay. The fault is again visible, forming a narrow cleft leading out to sea, from the Rotherslade pavilion. Horizontal slickensides, thick calcite and reddening (there is virtually a Triassic deposit here) are again features of this fracture.

Limestone horizons well below the 'Seminula Oolite' are visible in Langland Bay. Another light grey oolitic formation— the Caninia Oolite—is visible in Rotherslade Bay. This oolitic limestone is underlain by drab, fawn-coloured limestones—the Laminosa Dolomite—seen in central parts of Langland Bay and in old quarries just east of the large mansion which is now a mineworkers' holiday centre. Dolomite is a limestone rich also in magnesium carbonate. Above the Canina Oolite occurs a thin formation of alternating 'chinastones' and shales—the so-called 'Lagoon Phase' or 'Modiola Phase'. These beds are believed to have been deposited in widespread lagoons separated from the open sea of Carboniferous times by bars made of oolite sands. *Modiola* is a small fossil lamellibranch, shaped like a modern mussel. These lagoonal beds are visible on both the eastern (Rotherslade) and western shoulders of Langland Bay. A major anticline—the Langland Anticline—trends through the inner recesses of Langland Bay. Its axis runs through the golf course, whilst its *overturned* northern limb can be traced through the old quarries around the mineworkers' centre. This overturned

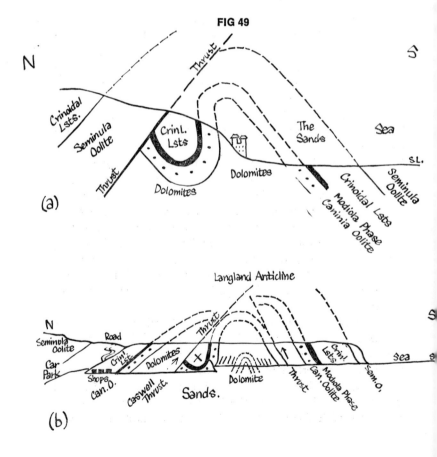

FIG 49

Langland upfold (Figure 49a) is further complicated by a power-ful thrust fault, best seen in Caswell Bay.

The long cliff stretches on either side of Langland Bay are skirted by a high-level coastal path. This path descends and rises into each minor inlet, incisions by the sea along minor faults. Also visible from this coastal path are large masses of angular limestone rubble called 'head'. This is a deposit dating back to the end of the Ice Age (about 11000 BC) caused by the sliding of thawed-out rocks over a still frozen foundation—a process called solifluction.

Caswell Bay Two inland-trending faults account for the siting of Caswell Bay. One fracture runs through the car park up Caswell Valley, the other occurs in the western inner corner of the bay. Both are 'tear' faults or 'wrench' faults, along which the slip movements were mainly horizontal. It has been proved that these horizontal slips occurred *whilst* the limestone beds were being folded. That faulting does occur in Caswell Bay can be demonstrated quite easily. For example, northerly dipping strata near the ice-cream shops, on the eastern side of the bay, are in line with southward dips on the far western side. That folding also occurs in Caswell can be seen on the eastern side (at about middle to low tide position) where there is a sharp syncline (**X** in Figure 49b) with a vertical southern limb. One can (after climbing up about six feet over the outer rock rim from the sands) walk around the western fringe of this syncline. The Caswell Thrust (mentioned above for Langland Bay) is also visible, just north of the syncline. This thrust, or reverse fault, dips northwards, the Dolomite having been pushed southwards over the (younger) Caninia Oolite during the folding of the rocks.

One can mention so many features of geological interest in Caswell. Suffice it therefore to dwell on but two other aspects. First, the sporadic deposits of beach shingle (with fossil shells) which now occur at about *thirty feet above modern sea level*. This is a Pleistocene Raised Beach, marking a higher sea level at some time in the Ice Age. This beach deposit is well seen on the extreme eastern coast of Caswell Bay (just north of Whiteshell Point), the shingle being overlain by boulder clay, showing that another ice advance occurred after the higher sea level. The occurrence of the limpet *Patella* (Figure 48) in the raised beach accounts for this deposit often being called the Patella Beach.

Secondly, the 'Modiola Phase' deposits occurring above the Caninia Oolite around the edge of the syncline (mentioned above) are worth observing, as they show evidences of having slid (and even faulted) whilst the deposits were being laid down. Moreover, the Caninia Oolite must have been eroded prior to the laying down of the basal shales and clays of the Modiola Phase, as is demonstrated by the very serrated and pot-holed top of the Caninia Oolite within the syncline. Many of the pot holes there are modern but some are obviously very much older (330 million years old as a matter of fact)!

THREE CLIFFS BAY

This deep inlet is named after the three-pinnacled crags on the seaward eastern side of the inlet (Figure 50) and owes its location to erosion along a great tear fault which runs from near Parkmill, past the valley side of Pennard Castle and out to sea. The rocks of the eastern side have been pushed northwards almost 200 yards relative to the western side of the fracture. As the limestone beds in Three Cliffs Bay are virtually vertical, one can demonstrate this relatively easily by matching up identified

FIG 50

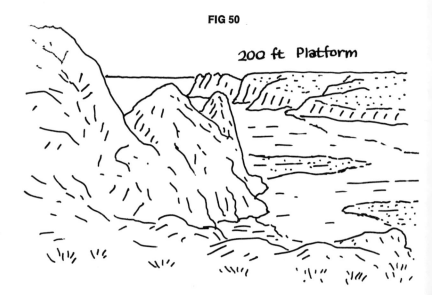

200 ft Platform

FIG 51

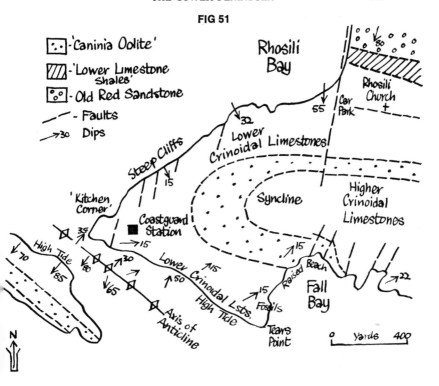

limestone formations on each side (particularly the Caninia Oolite).

The innermost rocks of Three Cliffs Bay reveal (imperfectly perhaps) the lowest formations of the Carboniferous Limestone together with the immediately underlying Devonian conglomerates. These very pebbly resistant beds at the top of the Old Red Sandstone are well seen on the craggy hill overlooking the holiday bungalows high up on the western part of the inlet. These quartz-conglomerates (with occasional pebbles of red jasper) dip southwards off the crest of the great Cefn Bryn Anticline. Their ridge-like outcrop is succeeded southwards, on this western side of the fault, by an E-W trending valley eroded into the softer Lower Limestone Shales (not very well exposed) which form the lowest formation of the Carboniferous Limestone. The sand dune spit running into the centre of the inlet is in line with this hollow.

Beyond the dunes, the main rock section on the western side of the bay begins. The succession here begins with well-bedded crinoidal limestones very rich in brachiopods. These fossil shells abound on the bedding planes, the most common genus being *Chonetes* (Figure 48). Some bedding planes are covered with worm burrows made by soft worms whose remains have never been preserved. Such worm tracks and burrows are usually termed 'fucoids' or 'trace-fossils'. Nodules of a siliceous flint-like substance called 'chert' can also be seen in the lowest beds (immediately seawards of the sand dunes).

An oblong-shaped shallow recess in this western cliff section shows some slight folding and faulting of the beds and brings on the Laminosa Dolomite (so called because a brachiopod *Tylothyris laminosa* has been found in such beds elsewhere). The dolomites give way seawards to the light grey Caninia Oolite, the Modiola Phase (marked by a very deep narrow cleft) and the overlying coarse, very crinoidal limestones. The latter form a small island, and are rich in gastropods, large single corals and the tabulate coral *Michelinia grandis* (Figure 48). Another fault occurs in this far western corner of the bay with again calcite, slickensiding and red iron oxide. The Seminula Oolite forms the *accessible* headland on this western side, but the coast turns westwards into Tor Bay and the Great Tor, a headland bringing on those higher limestones which were seen on Mumbles Head.

Before leaving Three Cliffs Bay reference should be made to Pennard Castle dominating the inland view from the sands of Three Cliffs Bay. The twin D-shaped gate towers and single arched gateway still mark the only entrance and date the present castle to the period of consolidation of Norman conquests in the later years of the thirteenth century. The present stone curtain walls stand on a rubble bank which enclosed and defended this site in the troubled twelfth century. The encroachment of sand over this eastern plateau of Three Cliffs and Pennard appears to have commenced soon after the stone castle was completed.

RHOSILI AND WORMS HEAD

This south-western tip of the Gower Peninsula is noted for its majestic sweep of Rhosili Bay (overlooked by the Old Red Sandstone ridge of Rhosili Down) and for the spectacular flat-topped Worms Head promontory. The latter is the southern limb of an important anticline. The northern limb is under the sea but the axis of the fold can be perfectly traced across the rocky causeway (exposed only at low tide) between Worms Head and the coastguard headland (see Figure 51). The tide comes in rapidly here, so take care! This upfold 'pitches' at both ends and is therefore an example of a 'pericline' (a structure shaped like an upturned boat). The core of the fold exposes those fossiliferous limestones seen just beyond the sand dunes in Three Cliffs. The Caninia Oolite comes on in Fall Bay (east of the coastguard station) but the underlying limestones here are not dolomitised. Many gash veins of red marl, fragmental 'breccia' and iron oxide in the inner recess of Fall Bay mark remnants of a one-time Triassic cover. The eastern boundary of the outer sweep of Fall Bay is marked by the Mewslade Fault, causing the deep inlet of that name. Fascinating modern beach springs ('sand volcanoes') occur here, beside wide calcite belts along the Mewslade fracture.

The village of Rhosili bears a name that is spelt in a number of different ways. One breakdown of the name is *Rhos*, meaning 'moor', and *Iley*, the name of a local stream. Its church has a Norman doorway believed to have been brought from a former church whose remains could have been seen a century ago in the Warren—the strip of meadowland between the beach and the foot of Rhosili Down. A memorial to Edgar Evans, Scott's companion to the South Pole, and a Gower man, is to be found within the church. The sands of Rhosili Bay are claimed to contain the wreck of a Spanish galleon. Quantities of silver coins were uncovered by the tide and found by villagers in 1807 and 1833. The ghost of one of the Mansel family, claimed to have seized the hulk of that cargo, is supposed to drive his coach and horses over the sands at midnight (the noise may, however, be readily explained by the sea driving through a blowhole on Worms Head). Other historical features of this unique village are its ancient open field system (strip-farming), its association with John Wesley, and the two megalithic tombs (the Sweyn's Houses) on Rhosili Down. They may mark the burial place of Sweyn, the Scandinavian sea lord who may have given his name to Swansea (Sweyn's-ey).

On the north side of the village the Lower Limestone Shales are involved in a zone of thrusting against the Old Red Sandstone quartz conglomerates of Rhosili Down. The shales (with tough dark limestone bands) are exposed near wet ground just beyond a farm gate. Brachiopods such as *Rhynconella* (*Camarotoechia*) and *Chonetes* occur in both the shales and limestones. Bryozoan lacework (*Fenestella*) can also be obtained. The Devonian conglomerates occur in an adjacent quarry. The foot of Rhosili Down is marked by the Broughton Fault (best seen in the bay of that name on the north coast of Gower). The sloping field-covered shelf at the base of the down is cut into hill-wash scree. The waters of Rhosili Bay may hide much younger (Triassic?) strata.

ARTHUR'S STONE

The return journey to Swansea across Gower from Rhosili can be made by road across the crest of Cefn Bryn, through the village of Reynoldston. The core of the Cefn Bryn Anticline exposes the reddish Devonian 'Brownstones' (beneath the quartz-conglomerates) in a small quarry just east of the summit parking area. At one time Silurian shales were claimed to outcrop along Cefn Bryn, but these strata have since been identified as the (Carboniferous) Lower Limestone Shales. Small outcrops of the conglomerates can be seen nearby. The structure of Cefn Bryn

FIG 52

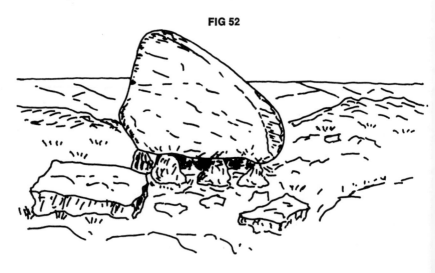

FIG 53

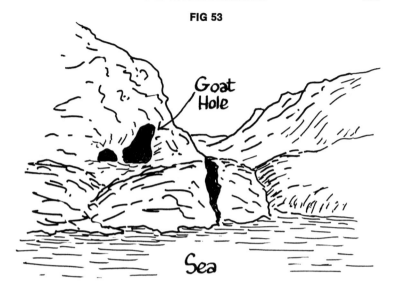

Goat Hole

Sea

is complex, with two anticlines separated by a downfold (marking the shales) and a thrust.

The shales account for the wet ground, with ponds, between the parking area and the remarkable Neolithic tomb known as Arthur's Stone. This can be reached by walking northwards along a wide grass track from the summit parking area. The large capstone is perched on a number of smaller plinth stones (Figure 52). Portions of the capstone have broken off (no one knows when) and lie near the plinth stones. The original capstone is said to weigh over twenty-five tons. The whole stone structure was probably the tomb of a chieftain, and was originally covered as a mound of earth. The dolmen is possibly the Maen Ceti, one of the wonders of the ancient Isle of Britain. Legend says that the capstone was a pebble in King Arthur's shoe! King Arthur's sword is said to have split the capstone.

The capstone was generally thought to be composed of the local (Old Red Sandstone) conglomerate. It has pebbles of quartz and even jasper. Of special interest, however, in the capstone (and its broken off portions) is the presence within the rock of pebbles of brown clay-ironstone and even one small fragment of coal has been noted. The author believes that the capstone is in fact formed of Millstone Grit conglomerate and comes from a formation on the *north* side of the South Wales Coalfield, in other words at least thirty miles away. Either then this twenty-five ton block of conglomerate was carried by Neolithic man, in about 2500 BC, to Cefn Bryn, a distance of at least thirty miles, or it was carried by an ice sheet to Gower and then man carried it a limited distance to its present site. The latter explanation (now also invoked for the Pembrokeshire stones on Stonehenge) would seem the more likely. Large, seven foot boulders of Carboniferous sandstone occur in the boulder clay of the north side of Rhosili Bay. Perhaps Arthur's Stone was carried to Cefn Bryn from there.

THE CAVES OF GOWER AND PAVILAND MAN

The limestone tract of southern Gower is noted for its numerous caves. Most of these occur on the southern cliffs but one inland cave system, Llethrid Swallet, contains fine stalactites and stalagmites. The best known of the southern sea caves are Paviland, Mitchin Hole and Bacon Hole. The two Paviland caves (Figure 53) occur along a remote cliff face, two miles west of Port Eynon. A signpost on the main road to Rhosili marks a path to the caves, but the caves can only be easily entered from the beach when tides are exceptionally low. Goats Hole, the more famous of the two caves, is now empty, but has yielded over eight hundred prehistoric implements and a wide variety of prehistoric animals. The most famous find, however, made in 1823, was of the headless left side of a human skeleton, called by Dean Buckland 'the Red Lady of Paviland' because he believed the skeleton (dyed with red ochre) to be that of a female. In 1913, Sollas proved it to be the remains of a Cro-Magnon youth. It is really the earliest ever found remains of prehistoric man, but unfortunately this was not realised for almost a hundred years. The skeleton, now in an Oxford museum, has recently been allotted a radiometric date of eighteen thousand years. Most people would have given it an older date. This Paviland cave has been used by man (and woman, as proved by such articles as necklaces made of periwinkle shells strung with animal gut) from Palaeolithic to Roman times.

Paviland Man is the only prehistoric human find in Gower. Many of the other Gower sea caves have, however, yielded human implements (even as old as the Mousterian culture) and the remains of Pleistocene animals. Mitchin Hole (or Minchin Hole), the largest of Gower's coastal caves, is situated in the cliff face 600 yards south-south-east of Southgate. This cave has been excavated on a number of occasions, even as far back as the beginning of this century. Remains of elephant, bison, soft-nosed rhinoceros and hyaena have been found. Most of these are now in the Swansea Museum. Remains of the 'Patella Beach' occur outside the cave. Bacon Hole, another well known bone cave, is situated further east along this stretch of coast. It showed evidence of occupation by man in Iron Age, Roman and later periods. The cave's name is derived from the red iron oxide streaks (of human or natural origin?) on the cave wall.

One last thought about Gower's caves; at the time of Palaeolithic man and probably up to the end of the Pleistocene Ice Age, the Bristol Channel was, at least in part, a wooded coastal flat. This was where prehistoric Gower Man hunted. His retreats were the rocky 'coastal' caves of the limestone tract of Gower. From here he could survey the lowland and see his enemies approaching. The rising sea levels of Mesolithic and Neolithic times gradually began to drown the Bristol Channel lowlands, slowly driving man inland in his search for food. The numerous Iron Age hill forts of North Gower and the mainland point to this eventual inward movement of early man in South Wales.

CHAPTER 7

The Swansea Valley

The Swansea Valley is one of the areas in Wales where the Welsh language is still fluently spoken and it can be said to have a character, and to some extent a dialect, that is all its own. Here the Welsh chapels and Sunday schools are still very much a part of life in the valley and the people take a pride in saying they are from the Swansea Valley. The valley has a long history of metal smelting and coal mining, the former going back over two hundred years. The city of Swansea has gained a great deal of character from the rich cultural hinterland of its valleys.

From scenic and geological standpoints too, the Swansea Valley is an extremely interesting area. The valley is drained by the River Tawe (the Welsh name for Swansea is *Abertawe*), and many tributaries enter the main valley, more especially on its northern side. These streams include the Lower Clydach (entering at the village of the same name), the Upper Clydach (at Pontardawe) and the Twrch (at Ystalyfera). This portion of the main valley between Ystalyfera and Clydach is narrow and deep (even deeper if one removes the superficial gravels and muds from the present-day floor of the valley). In fact, if these superficial deposits were removed and the solid rock floor revealed, then the sea would invade the valley to a depth of more than 100 feet and the 'inlet' would make an excellent glaciated fjord!

This portion of the Swansea Valley (from Ystradgynlais seawards) is cut mainly into the Upper Coal Measures (or Pennant Measures), predominantly rusty-weathering sandstones alternating with thinner shale sequences. The plateaux on either side of the valley reach heights of between 600 feet and 1,400 feet above sea level. Dominating the seaward end of the valley are the Pennant Hills of Townhill and Kilvey Hill, remnants of a 600 foot marine-cut surface (see Chapter 6). The higher plateaux overlooking the Clydach-Ystalyfera portion of the valley probably also mark eroded relics of yet higher platforms (at 800 feet, 1,000 feet and higher). Whether these higher surfaces were cut by marine action or are the result of long sustained planing down by rivers is a matter for discussion. Whatever the cause, these high-level platforms probably represent erosive phases of some two to eight million years ago, that is the Tertiary period known as the Pliocene. What makes this problem an interesting one is that platforms at comparable heights can be traced in many other hill districts of western and northern Britain.

The main axis of the downfold or basin forming the South Wales Coalfield lies in the vicinity of Morriston and Clydach. Thus the Pennant sandstones on Kilvey Hill (so clearly seen when one departs by train from Swansea's railway station) dip or tilt northwards whereas above Clydach they begin to dip back towards the south.

North of Ystradgynlais, the Coal Measures give way first to the underlying Millstone Grit and then to the Carboniferous Limestone which forms the striking serrated ridge of Cribarth, overlooking the hospital known as Craig-y-Nos ('rock of the night'), originally the home of Adelina Patti, the celebrated opera singer. The River Tawe cuts a gorge through the limestone ridge here. Nearby are the famous Dan-yr-Ogof caves, open to the public and renowned for their magnificent stalactites, stalagmites and rock curtains. These, and other underground caverns in the limestone on the opposite side of the valley (Ogof Ffynnon Ddu), are among some of the longest underground systems in Britain.

The base of the Carboniferous Limestone runs along the high right-bank wall of the steep valley known as Nant Haffes, and the long slopes north of this area are formed of Old Red Sandstone. These red Devonian grits and conglomerates form the majestic northward-facing scarps of Fan Gihirych (2,381 feet OD) on the east side of the Tawe's source and of Fan Hir, Bannau Brycheiniog (2,632 feet OD) and Bannau Sir Gaer (2,462 feet OD) on the western side of the watershed. The deep glacial-excavated lakes of Llyn-y-Fan Fawr and Llyn-y-Fan Fach nestle darkly at the foot of these corries or cirques. A famous Welsh legend is connected with Llyn-y-Fan Fach. It tells of a fair maiden who rose with her heirloom from the waters of the lake to marry a prince. If, however, the groom was eventually to strike the maiden three times with an iron bar then she would return with her possessions (cattle, etc) to the lake. As inevitably happens in a legend, the maiden was struck three times and returned to the depths of the lake. The sobs of the husband are supposed to be still heard around its shores.

The man-made lake north of the main watershed is the Cray reservoir which serves Swansea. The River Cray flows north-wards to join the Usk near the garrison town of Sennybridge. The River Senni, another feeder of the Usk, gives its name to the Senni Beds, red and green flagstones occurring beneath the main scarp-forming formation (the 'Brownstones') of the Old Red Sandstone. The Senni Beds are interesting in that they have yielded fossil fishes (*Pteraspis*) in Carmarthenshire and primitive fossil plants (*Gosslingia*) at a locality near the road running from

FIG 54

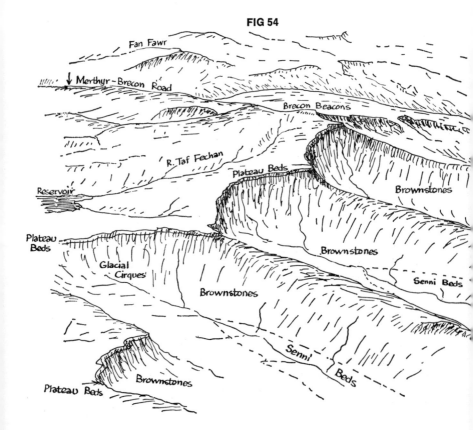

Merthyr Tydfil to Brecon (Grid Ref SN 971208). The Vale of Usk, from Sennybridge downstream to Brecon and Crickhowell, is cut into the thick fertile red marls which underlie the Senni Beds. The slopes from the summits of the Fans and the Beacons to the Usk Valley therefore give superb sections through the Old Red Sandstone, the main body of the scarp being held up by the very resistant Brownstones (red and brown sandstones) and their pebbly capping—the Plateau Beds (Figure 54).

ICE IN THE SWANSEA VALLEY

South Wales, like many parts of upland Britain, was extensively glaciated during the Pleistocene Ice Age. There were several ice advances and retreats in Britain and the last two advances (the Gipping and the Weichsel or Devensian) affected South Wales. During the Gipping glacial phase, ice caps formed on the Welsh uplands and glaciers moved down the major valleys and the same thing happened again in the last (Weichselian) advance. During the penultimate glacial phase, however, a major ice sheet moved down the Irish Sea and its ice fronts reached down to the Cornwall coasts and up the Bristol Channel almost to Cardiff. 'Foreign' (to South Wales, that is) boulders from Scotland, Ireland and the Lake District were dumped in boulder clays or 'tills' on the Welsh coasts, including such unique rocks as the Ailsa Craig microgranite (all the way from that island off the mouth of the Clyde in Scotland!). The local valley glaciers brought down boulders and pebbles of Old Red Sandstone, Millstone Grit and Coal Measure sandstones (limestone rarely occurs in the boulder clays) and dumped them in the valleys and even on the high interfluves between them. Scratches made by the moving, grinding ice have been noted high on the sides of the Neath and Swansea Valleys and as high as 1,700 feet OD in the headwater regions of the Tawe. Boulder clays were dumped by ice over Gower. Sections of boulder clay can be seen near the University Abbey in Singleton Park, Swansea and again in many road sections up the Swansea Valley. The boulders, large and small, are embedded in a stiff clay. You may well find a boulder which was scratched by the ice.

The local valley glaciers probably pushed out well into Swansea Bay but on the last retreat they began to melt back up the valleys to their ice cap sources. The melting appears to have been sporadic, with ice pauses at certain positions in the valleys. The debris dumped at each pause in the retreat of the glaciers forms long high mounds known as halt (or recessional) moraines. One of the best examples is that which almost blocks the Swansea Valley from the village of Glais across to near Ynystawe. This Glais Moraine (Figure 55) rises to 150 feet above the level of the river alluvium at its western end. Its steep southern edge faces a broad area of gravels washed out from the glacier snout by the streams of meltwater. This last ice retreat probably took place (on the basis of evidence elsewhere in Britain) about 10,000 years ago. Since then the climate has gradually ameliorated. It should be noted, however, that the milder 'interglacial' phases between the ice advances all lasted longer than 10,000 years so that we may now be living in yet another retreat phase with perhaps other glacial advances to come. Someone has, in fact, calculated on a 'cycle' basis that another glacial advance could happen in 70,000 years' time!

SWANSEA TO ABERCRAVE

The thick Pennant sandstones of Kilvey Hill give way northwards to the less resistant shales and thin sandstones of the highest Coal Measures in South Wales and which form the lower ground, near Morriston and Llansamlet, along the axial zone of the South Wales Coalfield basin. This low-lying area (overlain by glacial and river deposits and extending even from Hafod and Plasmarl) has been the location of metal-smelting industries in the Swansea Valley for almost two hundred years. In fact Chauncey Townsend founded a zinc-smelting works at Landore in 1757. Other zinc works opened in 1836 and 1841 at Llansamlet and Morriston. Ludwig Mond, a German, brought his nickel-smelting process to Clydach in 1900. In those earlier days Morriston and its surrounds was really the larger 'Swansea' but since then Swansea has spread southwards, and south-westwards into Gower.

The straight trench-like valley from Clydach to Ystradgynlais is eroded along a major, structurally disturbed belt in the Coal Measure (and underlying) strata. In that area the 'Swansea Valley Disturbance' is largely a narrow fracture zone, along which both

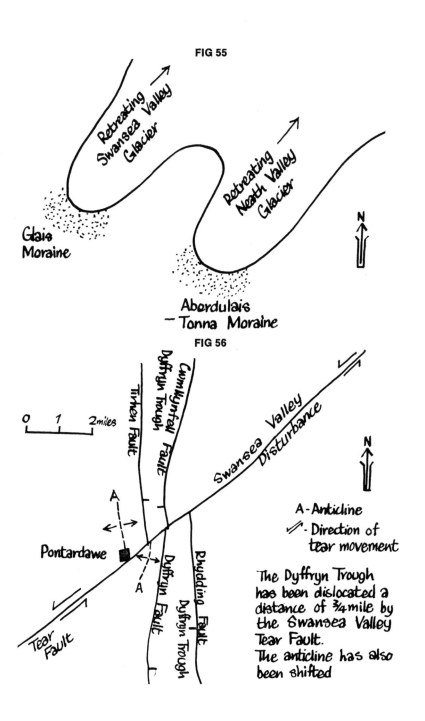

FIG 55

Retreating Swansea Valley Glacier

Retreating Neath Valley Glacier

Glais Moraine

Aberdulais — Tonna Moraine

N

FIG 56

Cwmllynfell Dyffryn Trough Fault

Tirben Fault

Swansea Valley Disturbance

0   1   2 miles

N

A
A
Pontardawe

Dyffryn Fault

Rhydding Fault

Dyffryn Trough

Tear Fault

A - Anticline

↗ - Direction of tear movement

The Dyffryn Trough has been dislocated a distance of ¾ mile by the Swansea Valley Tear Fault.
The anticline has also been shifted

horizontal (tear) and vertical movements have taken place in the geological past. The effect of this NE-SW aligned fracture-belt at Pontardawe is laterally to shift two faults trending across the valley in a south-south-eastward direction (Figure 56). Even a parallel anticline is displaced across the valley. This Swansea Valley Disturbance continues north-eastwards along the lime-stone ridge of Cribarth (where sharp folding becomes more important) and across the Old Red Sandstone escarpment (just south of Fan Gihirych) to beyond the town of Brecon. It prob-ably extends even to Shropshire (as a major fault along the south side of Titterstone Clee). Its south-westward continuation from Clydach is not clear but the author believes that it extends beneath the Llangyfelach area and the Burry Estuary.

This NE-SW belt of folding and fracturing in the Upper Palaeo-zoic strata probably marks an older underlying fracture-zone in the basement rocks (Precambrian?). Another similar weak zone in the uppermost crust lies beneath the parallel Vale of Neath Disturbance (Chapter 8). When the Devonian and Carboniferous cover rocks were compressed by the Armorican crustal forces at the close of Carboniferous times, slips occurred along the base-ment fractures, affecting the way in which the overlying strata buckled and fractured along these two NE-SW zones.

The present river pattern of the Swansea and Neath hinterlands (Figure 57b) is closely related to the location and trend of these two disturbed belts. The original river pattern of the area con-sisted of streams (like the two Clydachs, the Twrch and the Dulais) flowing southwards or south-south-eastwards. Thus the Upper Clydach (or Gors) once flowed across Pontardawe and Alltwen towards Neath Abbey. Similarly, the Gwys and the nearby Giedd flowed across Ystradgynlais and down the Dulais. This earlier Dulais may even have flowed still further southwards over the Bwlch Gap (three miles north-east of Port Talbot). Figure 57a shows this ancestral river pattern. The timing of the river captures is difficult to assess. One method is to note the heights of the abandoned channels or 'wind gaps'. The greater height of the Rhos Common Gap (512 feet), south of Ystradgyn-lais, as compared with that of the Alltwen Gap (300 feet) near Pontardawe, shows that a Giedd-Dulais river was captured by a 'Tawe' tributary of an Upper Clydach which was still flowing south-south-eastwards through the Alltwen Gap and which was

FIG 57

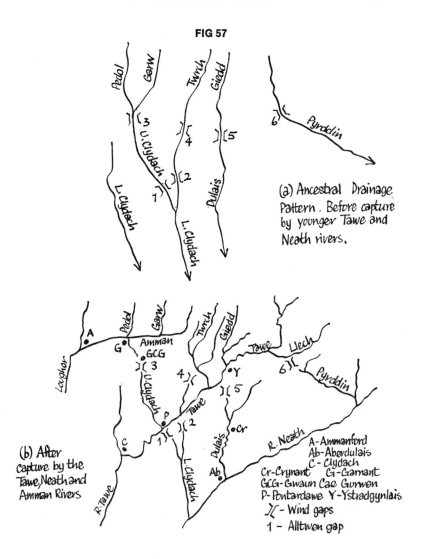

(a) Ancestral Drainage Pattern. Before capture by younger Tawe and Neath rivers.

(b) After capture by the Tawe, Neath and Amman Rivers

A - Ammanford
Ab - Aberdulais
C - Clydach
Cr - Crynant   G - Gamant
GCG - Gwaun Cae Gurwen
P - Pontardawe Y - Ystradgynlais
)(- Wind gaps
1 - Alltwen gap

graded to a sea level that was probably 200 feet higher than today. The 200 foot platform is now generally considered to date to one of the earlier interglacial phases of the Pleistocene, so we could say that the River Tawe had not yet completed its *full* course (as we know it today) some 300,000 years ago.

THE UPPER REACHES OF THE SWANSEA VALLEY

It is here that the scenery of the region attains a fascinating variety, with bare gritstone pavements (dotted with often large subsidence holes), sculptured grey limestone cliffs and crags, deep gorges and gullies with numerous rapids and waterfalls, and underground cave systems, one brilliantly illuminated and accessible to the public. The NE-SW trending ridge of Carboniferous Limestone, broken by the Tawe gorge into the Cribarth and Penwyllt portions (Figure 58) is particularly prominent. One

FIG 58

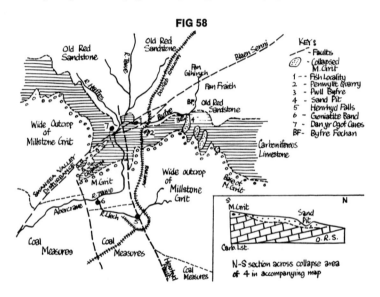

wonderful way in which to appreciate the scenic changes which occur as one travels northwards across the Millstone Grit and the Limestone on to the Old Red Sandstone is to walk northwards from Coelbren Station along the now disused railway to Craig-y-Nos Station and beyond. A magnificent panoramic view of the Cribarth limestone ridge will be one of the rewards. Moreover, one can imagine that one is travelling by railway carriage in the company of the fabulous Madame Patti and her gay entourage towards Craig-y-Nos Station on the way to her magnificent home where operatic airs would ring out into the night, mingling with the song of the nightingale in the nearby woods. A more recent 'imagination' has been to think of this rugged area as a South African vista during a recent filming of the life of a young Sir Winston Churchill.

Cribarth The Swansea Valley Disturbance takes the form of a complex anticline in this limestone area and the folds can be clearly seen on the Cribarth ridge. The crest can be reached either from Craig-y-Nos or via a turnstile-type gate and a steep winding path near the Abercrave Inn. The core of the main anticline is seen along the summit ridge and is cut into light grey oolitic limestone overlain by dark limestones containing much nodular silica in the form known as 'chert'. On both flanks of the ridge the blocky quartzites of the Millstone Grit form very rough boulder-strewn slopes. Between the uppermost limestones and the Grit on the north-western flank of the ridge is a marked hollow with many old trial holes. This is the thin outcrop of the 'Upper Limestone Shales'. These rocks often weather into a siliceous powder called 'Rottenstone', dug in older days for abrasives. In the days when brass ornaments adorned virtually every mantlepiece in the South Wales valleys there was a great demand for this polishing material.

FIG 59

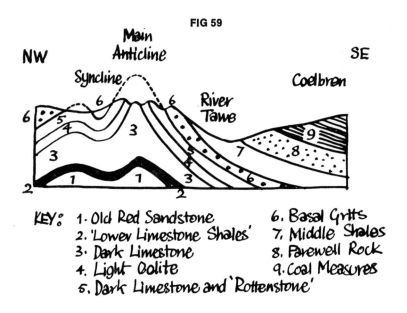

KEY:
1. Old Red Sandstone
2. 'Lower Limestone Shales'
3. Dark Limestone
4. Light Oolite
5. Dark Limestone and 'Rottenstone'
6. Basal Grits
7. Middle Shales
8. Farewell Rock
9. Coal Measures

Beyond the main Cribarth Anticline lies another minor upfold separated from the ridge by a well-marked minor syncline in the Millstone Grit quartzites. Figure 59 is a section drawn across the Cribarth ridge to show the detailed structure. The main fracture zone so characteristic of the Disturbance further down the Swansea Valley is not very obvious on Cribarth and it is likely that the main effect of the earth movements was to fold the massive beds of limestone rather than fracture them. The elusive fault belt may on the other hand run near to some old trial pits in crushed quartzite on the north-western side of Cribarth and could occur again near the farm called Garth, two miles south-west of the ridge. It is also difficult to trace the fracture zone through the Old Red Sandstone area east of the upper Swansea Valley and it probably splays out into a number of ragged fault lines, one of which separates the hills of Fan Gihirych and Fan Fraith and then continues north-eastwards into the Senni Valley, noted for the all-night vigils kept by its farmers to prevent preliminary surveys for a proposal to drown the valley as a reservoir. Another fracture must run near the main Swansea-Sennybridge road (A4067) as it approaches the source of the Tawe, for here the Brownstones dip steeply towards the road and stream.

Nant Llech   The River Llech, a left-bank tributary of the Tawe, flows in a deep narrow gorge cut into the basal Coal Measures and the underlying Millstone Grit. There is reason to believe that an ancestral Llech once flowed in the *opposite* direction as part of a long tributary to an early River Cynon. After capture by the Tawe, and the considerable deepening of that main valley by periodic falls of sea level and glaciation, the Llech has had to cut down rapidly along a relatively steep gradient in order to 'catch up' with the Tawe. Numerous waterfalls therefore occur, as for example near the farm called 'Melin Llech' ('the mill on the Llech'), where several large fossil tree trunks were found by Logan in 1856. The most spectacular waterfall, however, in the Llech Valley is the Henrhyd Waterfall, almost directly north of the village of Coelbren (Grid Ref SN 854119). This waterfall is caused by the Henrhyd Fault, a fracture crossing the stream at right angles and bringing Coal Measures (to the west) against the Farewell Rock of the Millstone Grit (upstream, to the east). A narrow gulley running up the right bank of the valley (just west

FIG 60

Spirifer
(Devonian - Carboniferous
brachiopod)

Lingula
(Long-ranged
brachiopod)

Productus
(Carboniferous
brachiopod)

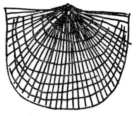

Dunbarella
(U. Carboniferous
lamellibranch)

Aviculopecten
(U. Carboniferous
lamellibranch)

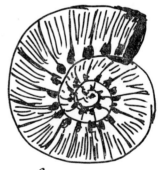

Gastrioceras
subcrenatum
(U. Carboniferous
goniatite)

Bothriolepis
(U. Devonian fish)
(Front dorsal portion
covered with bony
plates)

of the fall) marks the course of the fault and the face of the
fracture is seen to be highly polished and to bear grooves known
as 'slickensides'. The slickensides are vertical, showing that the
latest movements on this fault were vertical slips.

Downstream from the Henrhyd Fall are numerous exposures
of sandstones and shales. The latter are occasionally very dark
and yield small brachiopods, such as *Lingula* and *Productus*
(Figure 60), and shining fish scales. They are 'marine bands'
representing temporary incursions by the sea during the mainly
swampy conditions of Coal Measure times. A little tributary
entering the Llech below Cefn Byrle Farm reveals some three of
these marine intercalations, each forming a step in a rapid-
stepped stream profile. Just before its confluence with the Tawe
the Llech turns a right-angled bend to the right. A few yards
upstream from this bend is exposed the important marine band
which forms the base of the Farewell Rock—the *Gastrioceras
subcrenatum* Marine Band—and which in fact is the internationally agreed base of the Coal Measures in Europe. This
dark shale yields many examples of that goniatite (Figure 60)
plus brachiopods and lamellibranchs, the latter including
Dunbarella and *Aviculopecten*. Just downstream from the river
bend, a quartzite band is seen to be faulted against shale.

The Dan-yr-Ogof Caves   This extensive underground cave system (Figure 61) drains at least five square miles of Carboniferous
Limestone on the western side of the upper Tawe Valley. It
begins at a major 'sink hole' or 'swallow hole' known as Sink-y-
Giedd (height 1,436 feet OD). The mouth of the cave system, at
least two miles (in direct line) from its source, lies at a height of
770 feet OD.

The Morgan brothers were the first recorded explorers to enter
the dry series of caves above the river cave in 1912, and with
the aid of a coracle succeeded in crossing all but the fourth of
a string of lakes occurring some 400 yards in from the river cave
exit (and *just* beyond the limit of the show caves). The lamp they
used can still be seen near the entrance to the public caves.
Thirty-five years were to pass before this lake was conquered.
The public show caves were reopened in 1964.

In 1966 Eileen Davies succeeded in squeezing through into a
vast series of inner caverns containing well over two miles of
passages and large chambers and the most magnificent stalactites,

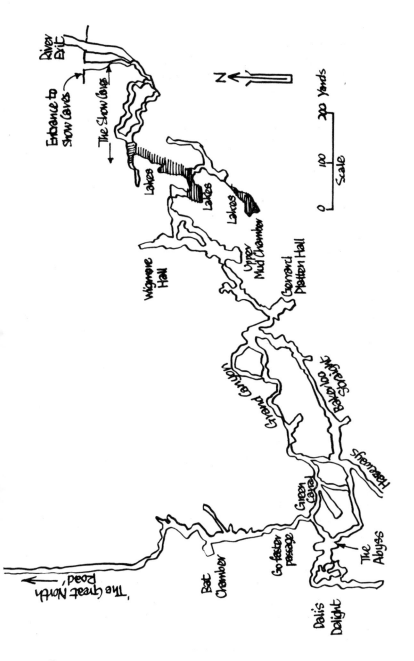

River Exit

Entrance to Show Caves

The Show Caves

N

100   200 Yards

Scale

0

Lakes

Lakes

Lakes

Upper Mud Chamber

Wigmore Hall

Clennard Phatten Hall

Bakerloo Junction

Green Canyon

Stalactites

Green Canal

The Abyss

Go faster passage

Bat Chamber

'The Great North Road'

Dali's Delight

E

stalagmites, flow sheets, curtains, etc. Individual stretches or caverns in this vast interior cave system now bear names such as Virgin Passage, Green Canal, The Abyss, Dali's Delight, Birthday Passage and Gerard Platten Hall (named after the organiser of the many exploring probes into the caves from 1947 onwards).

One feature of the show cave is the way in which it represents a *higher level* of drier caves above the modern river-traversed underground system. A fall of river level is thereby indicated and this 'double tier' system of passages is in fact a feature of other famous show caves—for example the Cheddar Caves. Presumably it results from a falling water table after the initial abundance of water consequent on a melting glacial cover.

Nant Byfre This left-bank tributary of the Tawe enters the main river near the housing estate just beyond the Caves car park. This tributary stream rises on the Old Red Sandstone but downstream flows on to the overlying Carboniferous Limestone. At least two of the fault splays along the Swansea Valley Disturbance cross the stream, one very close to the farm known as Pwll Coediog Farm (Figure 58). Permission should be sought at that farm to visit a fossil locality in the (Upper Devonian) Plateau Beds exposed in Nant Byfre high on its left bank some 300 yards east-north-east of the farm (and just beyond the N-S trending fence of the large field behind the farm). Here in the red flaggy sandstones and pebbly grits can be found plates of the Devonian fish *Bothriolepis* (Figure 60) together with the brachiopod *Spirifer* and calcareous seaweed (or 'algae'). The Devonian fishes adapted themselves to non-marine conditions, eg estuaries, lakes, river-basins, etc but this particular fossil layer must represent a brief incursion by the Devonian sea (which lay mostly to the south of Wales) into this region.

One last feature of geological interest in the Swansea Valley occurs just beyond the source of the Byfre. About one and a quarter miles east-north-east of the large limestone quarry at Penwyllt, another stream—the Byfre Fechan ('Fechan' means 'small')—disappears down a large amphitheatre-like sink hole known as Pwll Byfre ('Byfre Pit'). The dark limestone walls of this sink are over 50 feet high. This is where the Byfre Fechan disappears from sight to flow through the Ogof Ffynnon Ddu ('cave of the black spring') cave system and to resurge at Fynnon

Ddu (Grid Ref SN 846145)—near the River Tawe—a *direct* distance of over two miles. This cave system is probably the largest one known in the British Isles.

On the eastern fringe of Pwll Byfre is a Silica sand pit (Grid Ref SN 876176) from which sand was once obtained for making silica bricks at the Penwyllt brickworks. The deposit is known to have been at least 60 feet thick and is the northern end of a great tongue of Millstone Grit quartzites which periodically collapsed into underground solution cavities in the underlying limestone. The quartzites and pebbly grits have largely disintegrated into silica sand on collapse. The collapsed mass is almost 1,000 yards from north to south and its northern fringe rests on the basal shales of the Carboniferous Limestone, so that these northernmost quartzites have collapsed through a *total* height equivalent to almost the total thickness of the limestone, over 600 feet. One perhaps has to think of a long process of periodic smaller collapses, possibly over a long line of N-S aligned caverns, initiated possibly at the end of the Ice Age when the water table was higher. The present-day position of the sand is shown in Figure 58. Several other (almost as extensive) masses of collapsed quartzites occur further east along the limestone outcrop and there are numerous small swallow holes on the Basal Grit outcrop.

CHAPTER 8

The Vale of Neath

The River Neath crosses the central area of the South Wales Coalfield in a north-east to south-west direction. Like its neighbour the River Tawe, its main valley has been eroded along a major disturbed belt in the rocks—a narrow zone of quite intense folding and faulting. As with the Swansea Valley Disturbance, this 'Neath Disturbance' marks an important line of fracture or weakness in the underlying Precambrian foundation. Because of the disturbed character of the Coal Measures in the Neath Valley, no deep shafts are situated in the valley bottom. Levels, drifts or slants are instead driven into the valley sides and as a result the Vale of Neath has retained much of its original beauty.

This beauty is enhanced in the headwater region of the River Neath. There are a large number of headwater streams (the Nedd Fechan, Mellte, Pyrddin, Hepste, Sychryd, etc) and all abound in rapids and waterfalls. The Clungwyn Falls and the waterfall known as Scwd yr Eira are much visited and have been filmed on a number of occasions (often for stories unconnected with Wales). The deep character of the main Neath Valley has, furthermore, resulted in a large number of waterfalls on its 'side-tributaries', streams like the one entering the main river at Melincourt. (This waterfall is currently being filmed in connection with a film about King Arthur.)

Historically, the Vale of Neath has its interests. The mansion at Aberpergwm, near Glynneath, was the home of the important Williams family. The large white house near Cadoxton Church was once the home of H. M. Stanley (of Livingstone fame) whilst across the valley at Tonna there once lived Wallace, who put forward the theory of evolution at virtually the same time as Charles Darwin. Neath Abbey, situated to the south-west of

Neath, is an historic building. It was founded by the Norman lord, Richard de Granville, in 1129. Its history subsequently became more Welsh and it patronised Welsh learning and poetry. After 1530, however, it was taken over as a house by the Cromwell family and then later by the Herberts of Swansea, who built the Jacobean mansion part of the abbey.

The Neath Valley between Neath and Glynneath is deep and straight. The hillsides on Resolven Mountain and on Rheola Forest reach to nearly 1,300 feet above sea level. The valley floor, on the other hand, is below 150 feet OD almost up to Glynneath. The deep straight valley is due to the scouring action of a glacier which came down the Vale of Neath during the Pleistocene Ice Age (probably on at least two occasions). Ice scratches have been noted on rocks high up the valley sides, so that the valley was almost certainly filled to overflowing with ice. The ice irregularly gouged the valley bottom, deepening it appreciably in places. The irregular floor has been proved by boreholes which show that in places the solid rock floor is well below sea level. The deepening of the main valley by the ice action has steepened the profiles of the numerous tributary streams which now therefore 'hang' above the main valley floor, their waters rushing rapidly down the steep valley sides.

The Neath Glacier probably extended seawards into Swansea Bay, but its last retreat (about 11,000 years ago) was an irregular one, with pauses at about Briton Ferry, Tonna and Clyne. The recessional or halt moraines at Tonna and Clyne can be clearly seen (there is a cricket field on the former). The moraine material of the Briton Ferry halt is now only partially preserved and was probably originally only infillings between the rock ridges of this area (now utilised by the new bridge across the River Neath). Before the ice the river probably flowed more south-westwards across this region, through a now abandoned gap at Jersey Marine. When the Neath Glacier retreated as far as Tonna, the part-rock, part-moraine ridge near Briton Ferry held back the glacier meltwaters, forming a large 'Neath Lake' mainly to the immediate south-west of that town (Figure 62). This lake eventually spilled over near Briton Ferry, at the point now followed by the present river. It is likely that a second lake formed between Tonna and Clyne when the glacier had retreated to the Clyne position.

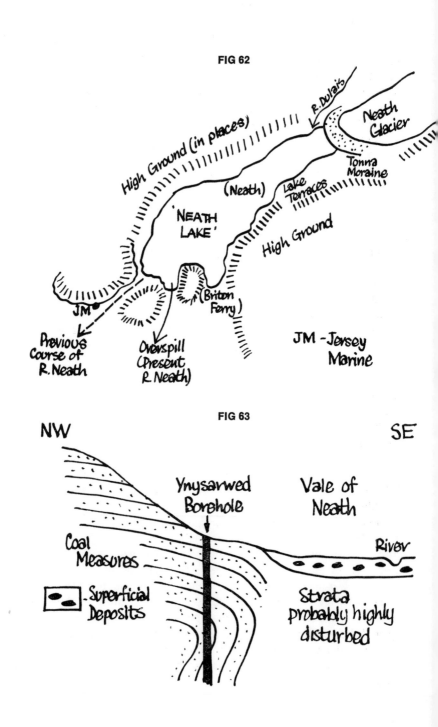

FIG 62

R. Dulais

Neath Glacier

High Ground (in places)

Tonna Moraine

(Neath)

Lake Terraces

'NEATH LAKE'

High Ground

JM

(Briton Ferry)

Previous Course of R. Neath

Overspill (Present R. Neath)

JM – Jersey Marine

FIG 63

NW

SE

Ynysarwed Borehole

Vale of Neath

Coal Measures

River

Superficial Deposits

Strata probably highly disturbed

Further evidences of ice action can be clearly seen on the high plateau of Craig y Llyn, overlooking Glynneath and the village of Rhigos. This plateau reaches almost to 2,000 feet above sea level and is a great capping of hard Pennant Sandstone. Into the steep northern and north-western slopes of this escarpment, the ice has carved several great amphitheatre-like 'corries' or 'cirques'. This has probably been the result of the frost shattering of the rocks on the underside of the thick ice, as it rested against the scarp slopes. The largest of these cirques surrounds Llyn Fawr reservoir (on the floor of which Iron Age implements were once found). The exciting climb along the A4061 over the escarpment from Rhigos and Hirwaun gives magnificent views down into the cirque and also northwards into the headwater region of the Neath and beyond to the Old Red Sandstone hills of the Brecon Beacons and the Breconshire Fans.

From this vantage point, one striking feature stands out in one's mind. At Rhigos (around the Trading Estate) the wide valley between Craig y Llyn and Penderyn Foel is dry. The River Sychryd drains the western end of it and enters the Neath river system, but there is no river draining right through the Rhigos gap. Looking away to the north-west, one sees the long course of the River Pyrddin, a course which appears to be a westward continuation of the gap at Rhigos. All this is explained by the evolution of the Neath-Cynon river systems. Figure 64 shows the probable stages in this evolution. Consequent on the falls of sea level in the Pliocene period, the ancestral river pattern began to form, with early rivers like the Cynon and Aman flowing in a NNW-SSE direction (stage 1 in Figure 64). Subsequently a major tributary to the Cynon developed, along a narrow belt of WNW-ESE faults. Adaption to joint patterns and belts of softer strata also changed parts of the courses of the headwaters (Mellte, etc). The last stage in the river evolution was the capture of the Cynon tributary by the new River Neath, developing rapidly along the Neath belt of disturbance (stage 3). The direction of flow of the River Sychryd was reversed and the wide valley at Rhigos became abandoned. The height of the Rhigos Gap (about 650 feet OD) suggests that the final capture of the Cynon's tributary took place when the sea level was 400 feet higher than today (the time of bevelling of the 400 foot platform in the Vale of Glamorgan—see Chapter 10). The rapid downcutting of the

FIG 64

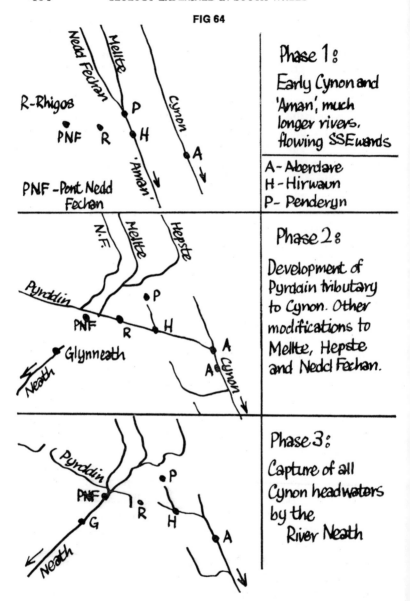

Phase 1:

Early Cynon and 'Aman', much longer rivers, flowing SSEwards

A- Aberdare
H - Hirwaun
P- Penderyn

Phase 2:

Development of Pyrddin tributary to Cynon. Other modifications to Mellte, Hepste and Nedd Fechan.

Phase 3:

Capture of all Cynon headwaters by the River Neath

new River Neath (re-enforced by its new-found headwaters) along its NE-SW disturbed belt plus the impetus given by the still-falling sea levels and the subsequent deepening glacial action has caused all the Neath's headwaters to 'hurry' their downcutting. This has resulted in the steep-sided narrow gorges along the Pyrddin, Nedd Fechan, Mellte, Hepste and Sychryd Rivers. These gorges are cut into the bottom of the broader, earlier valleys of these tributaries. This can be clearly realised by walking towards any one of these streams. Approaching them from some distance, one first descends gentle slopes (the earlier valley profile). Suddenly one comes to the sharp drop of the valley gorge, with the river rushing loudly below, and, moreover, often tumbling over rocky rapids or a lofty waterfall like Scwd yr Eira. As these rivers wear down their rocky floors, the rapids will be 'evened out' and the waterfalls will gradually retreat upstream and lessen in height. This will take time, however, and the magnificent grandeur of this Breconshire National Park is assured.

The River Neath and its headwaters give splendid exposures of the Devonian and Carboniferous rocks along the northern rim of the South Wales Coalfield. The Nedd Fechan, Mellte and Hepste cut deeply into the Carboniferous Limestone and the Millstone Grit. The waterfalls in the Grit are rivalled by the underground caves and dry valleys in the Limestone, the best example being Porth yr Ogof ('mouth of the cave') in the Mellte, near the village of Ystradfellte. From Glynneath to Neath, the solid rocks in the floor of the Neath Valley are hidden by the superficial veneer of glacial and river gravels, sands and clays. Nevertheless the shales, sandstones and occasional coals of the Lower and Middle Coal Measures are quite well exposed in the side-tributaries of the main river, for example in the streams coming in at Cwmgwrach and in the Pergwm and Rheola brooks. Other fine sections of the Coal Measures occur in Cwm Gwrelych, near Pontwalby. Most of these exposures in the Neath's tributaries show fairly gently-dipping strata and give little indication of the complex buckling and fracturing of the Coal Measure strata in the valley floor of the Neath. That the structure is complex, however, beneath the superficial cover in the main river was clearly demonstrated by a borehole put down at the village of Ynysarwed, near Resolven (Figure 63). At first, the strata penetrated were tilted only gently towards the valley, but at a depth of about 800 feet (at the horizon of the Yard Coal) the strata tilted up rapidly to vertical and even beyond. It is not surprising therefore that there is no deep shaft in this valley.

KING ARTHUR'S COUNTRY

Near the village of Pont Nedd Fechan, the various headwaters of the Neath come together to make the main River Neath from that point south-westwards. The village is the starting point for the glorious waterfall country. It is also near to the remarkable limestone ridge known as Craig y Ddinas (Figures 65 and 66). Besides revealing the intense folding and faulting along the Neath Disturbance, the ridge, about 200 feet high and extending in a WSW-ENE direction with sheer drops on either side (to the Mellte and Sychryd), is connected with a famous Welsh legend. A young man from West Wales was the seventh son of a seventh son and was thereby destined for great things. He left home to seek his fortune in England and on the way he was engaged to drive a herd of cattle to London. For his walking stick he cut for himself a piece of hazel growing on the sides of a rocky ridge. After delivering the herd he was stopped by a stranger who

FIG 65

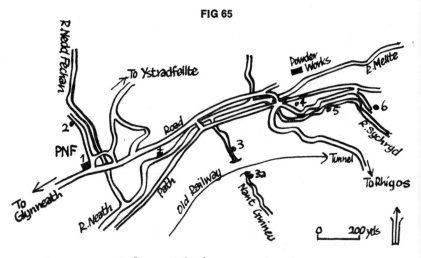

Localities: 1 - Farewell Rock   2 - G. subcrenatum
             3 - G. cumbriense   3a - G. cancellatum
             4 - Craig-y-Ddinas   5 - Bwa Maen
             6 - Dinas Silica Mine
             PNF - Pont Nedd Fechan

FIG 66

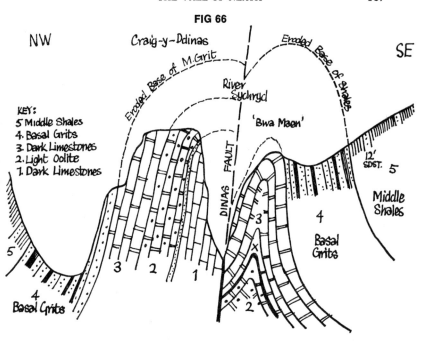

asked him where he had cut the hazel stick and that if he showed the way to the spot, then a vast treasure would be the reward. The ridge of the hazel tree was of course Craig y Ddinas. Digging up the foot of the tree they found a great flat stone which when lifted revealed an underground passage. Following the passage they came into a great cave, filled with warriors in bright armour, all asleep on the floor. In the middle lay a king and near him were two tall heaps of gold and silver. The young man was informed that he could carry out a pile of either gold or silver but that he was to watch that he did not disturb the sleeping warriors. To cut a long story short, the youth's greed led him back to the cave several times and on the last occasion he woke up the whole warrior group. He was thrown out by an angry King Arthur (for it was he) and when he later picked up enough courage to return, he could not find the stone. It is today alleged, however, that the cavern is inside Craig y Ddinas and that King Arthur and his knights still sleep there. When the Cymry (the Welsh) are in great danger, then the king will come forth to save them.

The entrance to this underground cave system is visible on one's left-hand side as one climbs up a wooden staircase over a deep cleft in the River Sychryd between Craig y Ddinas and the anticline known as Bwa Maen ('the bow of stone'). To get to this anticline and the staircase (which once was a route for trucks loaded with quartzite from the Dinas silica mine), follow the road which skirts the right bank of the Mellte from Pont Nedd Fechan. Crossing the Mellte by a bridge (just above the confluence of the Mellte and Sychryd), take the path round to the half left and follow the right bank of the Sychryd. This brings one to Bwa Maen and the foot of the staircase. The anticline is a fine example of the folding along the Disturbance. Note that the south-east limb of the fold is practically vertical. The limestone of the fold is dark in colour and contains thin seams of clay and even coal smuts. One such clay seam can be seen low down in the base of the fold. It is in fact likely that the limestone beds rumpled more easily above these lubricated clay seams and that the fold dies out with depth.

On the south-east side of Bwa Maen the limestone gives way to the steeply dipping Basal Grits of the Millstone Grit. This old quarry was made in the last century. The almost pure quartzites are used to make silica or refractory bricks, for lining steel furnaces, but none is now obtained from this Craig y Ddinas area. A little further up the Sychryd (after climbing the staircase and proceeding along the path along the right bank of the stream) one comes to the last area to be worked for refractory quartzite —the famous Dinas Silica Mine. This must have been one of the few areas in the world where quartzite was mined underground. One bed of quartzite was followed over an area of about 1,000 by 500 yards. Pillars of quartzite were occasionally left standing and there were many galleries, all well illuminated. The remarkable thing was that in an area of great structural complexity, the biggest fault throw encountered underground was 12 feet. Yet nearby is the great Dinas Fault, the great tear fault of the Neath Disturbance. It runs up the bed of the River Sychryd, right under the wooden staircase. As one climbs the staircase, a glance down into the stream bed will see the shattered nature of the limestone, with many veins of calcite. To the left of the gorge rises the ridge of Craig y Ddinas. It is not surprising that it is called 'the City Rock', knowing that King Arthur is sleeping inside.

From the Silica Mine one can climb a narrow path leading eventually to the flat grassy top of Craig y Ddinas. From here, fine views down the Vale of Neath can be obtained. Moreover, in low rocky crags of white oolitic limestone (with an internal structure like the roe of fish), one can obtain many good specimens of the brachiopod *Productus hemisphaericus*. The way down to the bridge over the Mellte lies down a rocky path. The bed of the Mellte lies far below and on its right bank are the ruins of the Nobel Powder Works. This works manufactured explosives for use in quarries and mines. It began operating in 1858, when water from the Mellte was used for power and the woods along the Mellte provided the necessary charcoal. It ceased operating in 1931.

Before leaving the Pont Nedd Fechan area, attention should be drawn to two other localities, both rich in Millstone Grit fossils. One locality lies a little way up the stream called Nant Gwineu, which enters the Mellte (on its left bank) 400 yards below the Mellte bridge (and just upstream from a small footbridge over the same river). About 150 yards up Nant Gwineu from the riverside path is a narrow cleft along that tributary, with an exposure of dark shale below a light-coloured quartzite. The shale is very fossiliferous and is the *Gastrioceras cumbriense* Marine Band (representing a time in the Millstone Grit when muddy seas temporarily spread over South Wales deltas). Besides the characteristic goniatite (see Figure 67), there also occur the brachiopods *Schizophoria* and *Productus* and the lamellibranchs *Nucula* and *Aviculopecten*. The new 'Heads of the valley' road was, at the time of writing, being constructed very near to this exposure, and new exposures of the fossil band could therefore emerge. Still further upstream (just beyond the position of the old railway line) is an exposure in the stream of another dark marine band in the Middle Shales of the Millstone Grit. This contains the goniatite *Gastrioceras cancellatum* (Figure 67) plus the lamellibranch *Dunbarella* (Figure 60).

The other locality is the region behind the Angel Hotel in Pont Nedd Fechan. Immediately behind the inn is an exposure of the Farewell Rock, the sandstone formation which occurs at the top of the Millstone Grit. At the south-western end of this sandstone bluff, a small anticline occurs in shales interbedded in the main sandstone beds. Moreover, the main sandy bed

FIG 67

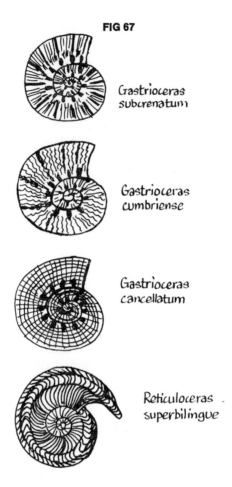

Gastrioceras
subcrenatum

Gastrioceras
cumbriense

Gastrioceras
cancellatum

Reticuloceras
superbilingue

thickens appreciably away from this fold and is seen to contain marked grooves (Figure 68). All these features are believed to be associated with the sliding of the sandstone beds when the latter were still wet and in a soft, plastic state. The sliding may well have been initiated by early movements, causing earthquake shudders, along the Dinas Fault. A brief walk up the valley behind the hotel reveals the full succession of the Farewell Rock which is then directly underlain by dark shale, containing *Gastrioceras subcrenatum*.

FIG 68

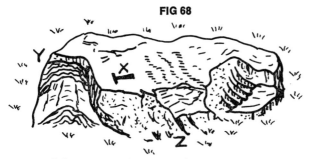

At Y the sandstone edge is only 4 in thick
At Z it is 5 ft thick
X is a hammer to give scale

THE WATERFALLS

Many waterfalls occur in the Neath's headwater streams and especially in the Pyrddin, Mellte and Hepste. These falls occur in the Millstone Grit area and are due partly to the alternation of hard quartzitic grits with softer bands of shale. The overall gradient of these streams is fairly steep because of the over-deepening of the main River Neath, for reasons stated earlier. Falls can further be caused by the sharp bringing together of hard and soft strata by faults and there are a number of these in the head of the Neath Valley. In some cases, for example Scwd yr Eira, waterfalls originally began at the fault position but have since retreated upstream. A well-written description of the falls has been given by the late Dr F. J. North (see reference, p 199).

Scwd Gwladys This fine waterfall occurs on the lower part of the River Pyrddin, just above its confluence with the Nedd Fechan. To reach it, continue up the path, behind the Angel Hotel, along the right bank of the Nedd Fechan. At the confluence of that river with the Pyrddin one can cross by a bridge to the left bank of the Pyrddin and a brief gradual climb brings one to a big rock perched high on the river bank and known as the Logan Stone. The top of the waterfall is near. The fall (Figure 69) is caused by the ledge of resistant quartzite, known as the Twelve Foot Sandstone, low down in the Middle Shales. The

FIG 69

Scŵd Gwladys
(on the R. Pyrddin)

Scŵd yr Eira
(On the R. Hepste)

underlying dark shales are marine in origin and contain the goniatite *Reticuloceras superbilingue* (Figure 67). Some undercutting of the shales has occurred, due to the spray, but one cannot yet walk under the water, as in the case of Scwd yr Eira.

Scwd Einion Gam Further upstream is a narrower fall. 'Scwd' from 'Ysgwd' (meaning 'toss' or 'fling') and 'gam' (meaning 'crooked') are appropriate descriptions of this narrow, twisting cleft down which the water roars in very wet weather. The fall is caused by a fault which has an upstream downthrow, to bring soft Middle Shales against the Farewell Rock. The ledge at the base of the fall is due to a hard band of quartzite in the Shales. This fall is difficult to reach from Scwd Gwladys. It is in fact easier to reach it from the road to Banwen, though the going can be wet on occasions.

Scwd Clungwyn (Upper Clungwyn Fall) There are really three Clungwyn falls in the River Mellte. This is the uppermost and occurs where the important Clungwyn Fault crosses the river. The fracture had a complex geological history, with both horizontal and vertical movements having occurred along it. In very dry weather, when the river is low, one can see spectacular 'slickensides' (grooves) on the fault face—this is also the face of the waterfall. The throw or displacement of the fault is about thirty feet, the same as the height of the waterfall. The river upstream flows along the top of the Basal Grits, jumps over the eroded fault face and regains the same (downthrown) bed of grit on the down-river side (Figure 70). The displacement can be worked out in another way too. The Basal Grits of the main fall mass are overlain by about twenty feet of shale (the shale of Scwd Gwladys) and thence by the Twelve Foot Sandstone. On the down-fall side that Sandstone can be seen to be brought by the fault against the top of the Basal Grits on the upstream side.

Further downstream are the other two falls. The middle one (Scwd Isaf Clungwyn) is a complex, threefold, twisting fall, described vividly by Richard Warner in 1813. The description by William Young in 1835 includes the words:

> Now right, now left, its devious course descends;
> In grandeur wild, the broken waters flow.

The fall is over ledges of Basal Grit, the river trying its best to get over a strip of these tough beds, caught between two faults, as rapidly as possible.

FIG 70

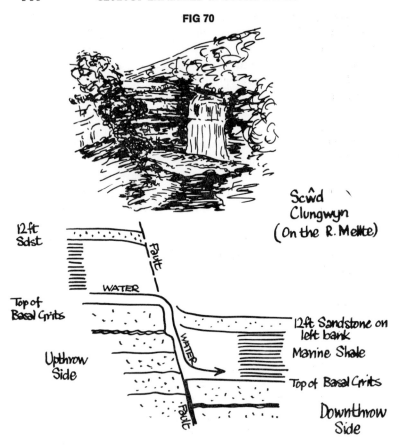

Scŵd
Clungwyn
(On the R. Mellte)

12ft
Sdst

Fault

WATER

Top of
Basal Grits

WATER

12ft Sandstone on
left bank

Marine Shale

Upthrow
Side

Top of Basal Grits

Fault

Downthrow
Side

Scwd yr Eira   This (Figure 69) is perhaps the most popular, being the one that people can walk under. This fall is in the River Hepste, the tributary of the Mellte. Upstream, the Hepste flows over Basal Grits for a considerable distance. At the waterfall it jumps over the topmost beds of the grits to reach the main mass of grits below. In between is a shale horizon (containing gonia-tites) which now forms the recess along which one can walk under the cascading water. The spray has eaten back this recess into the softer shale band, leaving the overhanging grit beds above to 'push out' or 'toss' the water (hence 'scwd'). The word 'eira' (meaning 'snow') appears to be a later corruption of the word 'eirw' (meaning 'cascade'), although in very cold, icy weather, the later term is often appropriate, with the fall frozen up. Just a short distance downstream from the fall is a narrow side-ravine or gully (on the left bank of the Hepste). This gully is in fact the best approach to the waterfall from the direction of the village of Penderyn (there is a large boulder—a glacial erratic—at the top of the gully: the roar of the fall will guide you from here). The gully marks the course of a fault with a downthrow upstream (the Twelve Foot Sandstone and the top of the Basal Grits can be picked out on either side of the gully as one descends). The waterfall must have originated at the place where this fracture crosses the Hepste, but it must have happened when the stream bed was at the position of the top of the Twelve Foot Sandstone on the upthrow side of the fault, with the shales which now cause the fall's recess being in the stream bed on the downthrow side. How long it has taken the river to erode back the waterfall upstream to its present distance away from the fault is difficult to assess.

UNDERGROUND RIVERS

One other feature of the Millstone Grit country in the Neath headwater region is the large number of deep holes in the land surface. These can be especially well developed near the base of the Millstone Grit, as on Cefn Cadlan (Penderyn).

Many small streamlets disappear down large swallow holes on the Millstone Grit moorlands around the villages of Penderyn and Ystradfellte. In the Nedd Fechan, Mellte and Hepste, however, bigger examples of underground drainage occur. In the first mentioned river there is the river cave of Pwll y Rhyd ('Pit of the stream'). In the Hepste, the river flow is mostly below ground for some considerable distance both above and below the bridge that carries the road from Ystradfellte to Penderyn. It is weird standing on that bridge and seeing a completely dry boulder bed below. Grass and rushes are growing on the stream bed upstream. If one puts an ear down to the stream bed one can hear water flowing below.

Porth yr Ogof The best-known example in the district is, however, the disappearance underground of the River Mellte at a point just over half a mile due south of Ystradfellte. There is a Youth Hostel nearby and from that hostel one takes a very narrow road (all right for cars if one blows the horn) off to the valley. The cave can also be approached by a narrow road from the east. Neither of these two approaches are suitable for a coach, but for cars there is a car park now. 'Porth yr Ogof' means 'mouth of the cave' and is where the Mellte goes completely underground for just over a furlong. One can walk (through a gate) over the dry stream bed to see the river emerge downstream. One word of warning: the cave system has its risks for the novice and there is an underground lake that presents dangers.

The cave has been described by many early travellers. Writing in 1698, Edward Lhuyd (one-time Keeper of the Ashmolean Museum in Oxford) mentions the *noted* cave called Porth Gogo at Ystrad Vellte in Brecknockshire and talks of the 'cockle shells' in the limestone. In 1809, Theophilus Jones talks about the famous White Horse (a coating of calcium carbonate on a wall at the end of the main frontal cave; it looks like a horse with a long neck). He wrote of it as 'a vein of calcareous spar supposed to resemble a naked child standing upon a pedestal, from whence (the pool beyond which it stands) is called Llyn-y-baban' ('the lake of the baby').

The disappearance of the Mellte underground is due to the differing character of the various units of the Carboniferous Limestone. About fifty feet below the top of the Limestone, a massive light oolitic formation occurs. This oolite is not as closely jointed (in the Ystradfellte area) as the underlying dark, blue-grey limestones and it is this contrast in the jointing that has decided the site of Porth yr Ogof. At the main entrance (Figure 71), the roof of the cave is of the massive oolite, whereas the river flows on the very closely jointed dark limestones. The ledges on which one walks to get into the cave are also of the darker formation. If one makes one's way carefully into the cave for some thirty to forty yards, a view of the White Horse is obtained.

The initial disappearance of water underground must have occurred at the point marked X (Figure 71a) well upstream from the cave, where the oolite first appeared in the stream bed. At first, some water must have made its way down joints at that point whilst the remaining water flowed downstream along the surface valley. In time, more and more water took an underground

FIG 71

Porth yr Ogof

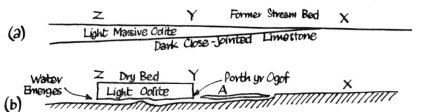

course and gradually also the collapse of a slender roof began to take place at the initial point of disappearance. The collapse has subsequently proceeded as far as Porth yr Ogof, as far as the main original surface bed is concerned. A lot more water is flowing underground in the broad area north of the cave, however: if one walks upstream from the descent path to the cave, a low sunken cavern is seen on the left side of the valley and a considerable volume can be seen to be flowing underground at this point.

Returning to the road above the cave, and proceeding through a gate, one can continue down the valley along the now dry river bed. Polished low rock cliffs tell of the earlier surface river. The roof of the cave system has collapsed in two places. Eventually one comes to the exit of the Mellte through a narrow gorge. One can then proceed further downstream to the Clungwyn Waterfall, crossing over a bridge before the fall.

THE PENNANT SANDSTONE

The Pennant Sandstone occurs in the upper half of the Coal Measures of South Wales and is responsible for the high hills and extensive high plateaux within the heart of the coalfield (see Chapter 9). In the Vale of Neath these tough blue-grey but rusty-weathering sandstone beds are responsible for the high ground on either side of the valley from Glynneath down to Neath, Briton Ferry and Jersey Marine. It is worth examining these Pennant strata because recent sedimentological studies have revealed much about their mode of origin, some 290 million years ago. The Pennant sands, muds and gravels were transported by rivers which flowed into the area that is now South Wales. The majority of these streams flowed off a mountain area situated somewhere to the south of Wales, possibly as far away as Devon and Cornwall. The rocks of this mountain area were eroded rapidly and the erosive spoil carried into the great fluvial basin that was South Wales. The further they flowed the more they began to swing and meander. These rivers cut channels into the underlying muds and sands. It has been possible to locate many of these 'fossil river valleys' by examining closely the Pennant rocks in quarries (and it is a much quarried rock, for building) and in stream sections and new road cuttings. The features to notice can be demonstrated with reference to sections near the new bridge (over the River Neath) between Jersey Marine and Briton Ferry. Almost beneath this bridge, very near to an inn, is a large disused quarry in the Pennant Sandstone (Figure 72b). The beds dip northwards and a considerable thickness of Pennant is exposed. The south face of the quarry shows a thick fireclay or 'seat-earth', underlying coaly films. This fireclay is the hardened product of the muddy slime in which the Coal Measure forests grew. Many logs and roots can be plainly seen in the deposit. It is the northern face, however, which is

FIG 72

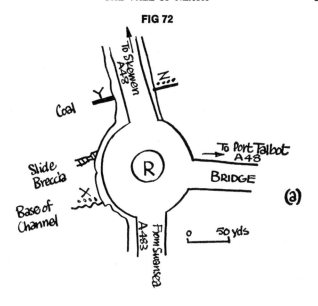

(a)

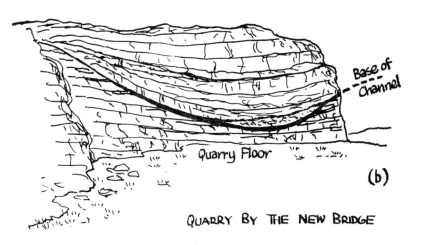

QUARRY BY THE NEW BRIDGE

(b)

of the greatest interest for this shows the base of a major
'channel', marking the base of a Pennant river valley. The base
of the channel decends to within a short distance of the base of
the quarry face. The basal band of the channel is of irregularly

laminated shale with stringers of sand, carbonaceous smuts and abundant plant fragments, some large. Near the eastern end of this quarry face the shale is replaced by a wedge of massive sandstone, up to ten feet thick, carrying large scattered fragments of coal, ironstone and plant debris. The sandstone has eroded the underlying beds and marked grooves can be seen on the bottom of the sandstone, showing how it has scoured into the underlying strata. The orientation of the grooves and of drifted fossil logs show that the fossil river was flowing from the south-east towards the north-west. The interesting aspect of this is that there was a river in this vicinity almost 300 million years ago, just as there is today, but the ancient river flowed in the opposite direction. Another interesting feature of the channel beds is the presence of coal fragments. These were carried by this Pennant river and this means that coal seams somewhere to the south of Wales were being eroded at the time. Were the fragments actually coal at that time or were they in a more woody condition? This problem is not unconnected with the problem of the origin of Anthracite coal (see Chapter 4).

Near this quarry is the Earlswood roundabout (Figure 72a) at the western end of the new bridge. Good sections can be examined here in the Pennant Measures. Of particular interest is the section on the west face of the roundabout (**X** to **Y** in Figure 72a). Here is exposed a complete cycle showing the erosion of a river channel and its ultimate filling by river-carried sediment. On the whole the sediment tends to become finer upwards within the cycle. The base of the river channel is seen at **X** to contain numerous flattened casts of logs, up to nine inches in diameter. Above this base are massive sandstones with large-scale cross-bedding. Above these beds is a 'slide-breccia' of crumpled shales, wedges of sandstone and plant casts, all in a chaotic arrangement. The top of the fluvial cycle is at **Y**, in the entrance of the road to Skewen from the roundabout. The six-inch coal overlies a four-foot fireclay, the coal swelling appreciably high in the cliff. The channel had by this time filled with sediment and a forest had become established for a while over a broader area of the basin. The same coal position can be seen on the opposite side of that roadway, at **Z** in Figure 72a, but here the erosive base of the next river channel can be seen.

CHAPTER 9

The Coalfield Interior

The main South Wales Coalfield is bisected into two portions by the Vale of Neath. The western portion is much lower in general height and has been described already in Chapter 4. In this chapter it is therefore proposed to describe the eastern portion—a much higher, rugged area, dissected by some fifteen deep valleys all noted in the past for their coal-mining activities. This area is, in fact, what most people think of as typical South Wales —high moorland plateaux entrenched by deep valleys along which run long parallel lines of houses. The pit head shafts (now largely disused) form landmarks in each valley. Also now to a large part disused are the snaking railway tracks which appeared almost as if from nowhere in each narrow valley in the second half of the last century. The region has been one of great exports —coal, steel, teachers, preachers, famous rugby players, choirs and actors. It has been an area of 'firsts'. The first railway— Trevethick's—ran from Abercynon to Merthyr Tydfil in 1804. The first cremation (by Dr William Price of his little son in 1883) took place at Llantrisant, near Pontypridd. A great Welsh athlete —Guto Nyth Bran—ran a famous race near Mountain Ash in the same century.

Geologically the area is a great basin. On its northern side, beyond Aberdare, Merthyr Tydfil and Ebbw Vale, lie the east-west outcrops of Carboniferous Limestone and Millstone Grit (the Heads of the Valley road runs practically along the base of the Coal Measures—in many places following the now disused railway line). On the southern side of the area lie the complementary 'South Crop' outcrops of limestone and grit, seen for example at Taff's Well, Machen and Risca (Figures 73 and 74). On the eastern side, too, the limestone and grit form narrow outer rims to the coalfield, as at Pontypool and on the prominent mountain known as Blorenge.

FIG 73

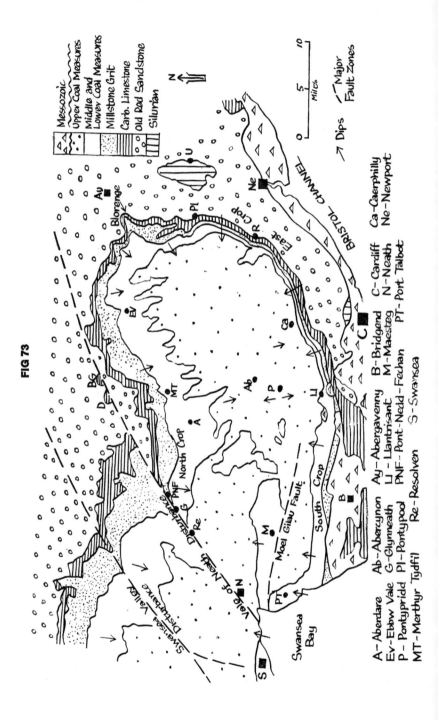

Messozoic
Upper Coal Measures
Middle and
Lower Coal Measures
Millstone Grit
Carb. Limestone
Old Red Sandstone
Silurian

Major Fault Zones
Dips
Miles

BRISTOL CHANNEL

East Crop
North Crop
South Crop
Moel Gilau Fault
Vale of Neath
Disturbance
Neath Disturbance
Swansea Valley Disturbance
Swansea Bay

A- Aberdare   Ab- Abercynon   Ay- Abergavenny   B- Bridgend   C- Cardiff   Ca- Caerphilly
Ev- Ebbw Vale   G- Glynneath   Ll- Llantrisant   M- Maesteg   N- Neath   Ne- Newport
P- Pontypridd   Pl- Pontypool   PNF- Pont-Nedd-Fechan   PT- Port Talbot
MT- Merthyr Tydfil   Re- Resolven   S- Swansea

FIG 74

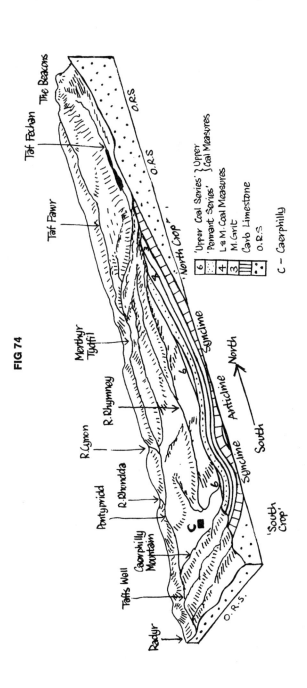

Radyr

Taffs Well

Caerphilly Mountain

Pontypridd

R. Dhondda

R. Cynon

R. Rhymney

Merthyr Tydfil

Taf Fawr

Taf Fechan

The Beacons

O.R.S

O.R.S

O.R.S

O.R.S.

North Crop

'South Crop'

Syncline

Anticline

Syncline

North

South

C

6 'Upper Coal Series' } Upper
4 'Pennant Series'   } Coal Measures
    L & M. Coal Measures
3 M. Grit
    Carb Limestone
    O.R.S.

C — Caerphilly

Within these curved bounding rims of Carboniferous Lime-stone and Millstone Grit lie the Coal Measures, up to 7,000 feet of shales, sandstones, coal seams and fireclays. The Lower and Middle Coal Measures are predominantly shaley successions, with many good workable coals—of bituminous (gas and coking) type in the east and south-east and of steam grade in the Rhondda, Ogwr, Cynon and Taff areas. The high quality steam-raising coals of those valleys led to the rapid development of mining in the last century when the steam locomotive was Britain's main mode of transport power.

The Upper Coal Measures, on the other hand, are formed mainly of resistant Pennant Sandstones, alternating with thin shales and some coals. It is these tough sandstones that give rise to the high plateaux of the coalfield interior. The Rhondda rivers rise in the Craig-y-Llyn plateau, almost 2,000 feet above sea level. The Garw and Ogwr Fawr rise in ground over 1,800 feet OD. The high 'backbones' between each valley, from the Cynon eastwards, reach heights of between 1,500 and 1,900 feet above sea level. The valley sides are scarred with old Pennant Sandstone quarries, the grey (rusty-weathering) stone having been in great demand for the building of the thousands of houses in each valley. In several places huge curved scars in the Pennant rocks mark the sites of former landslides. These are especially common in the Rhondda Fawr Valley and are believed to date back to the close of the Pleistocene Ice Age. Other marked Pennant scars are the numerous cirque-amphitheatres at the head of the Rhondda Fawr, Ogwr Fawr and Garw Valleys. The road journey from Port Talbot up the Avan Valley and over the watershed to Treorchy in the Rhondda Valley is a panoramic 'must' with spectacular hairpin bends in the road as it winds round the glacial cirques. So also is the road journey over Craig-y-Llyn, via the majestic Llyn Fawr cirque, from Hirwaun to Treorchy. The wonderful panorama from the road summit on Craig-y-Llyn (spanning all the Old Red Sandstone hills from the Brecon Beacons westwards) is just reward for the climb to the plateau top.

The workable coal seams of the region lie mainly in the shaley Lower and Middle Coal Measures. At the northern ends of the Taff, Cynon, Rhymney and Ebbw Valleys these beds are exposed on the surface and the coals have been frequently worked by 'levels' or 'drifts', that is, relatively flat passageways into the hillsides. Further down these valleys, however, and in more enclosed valleys such as the Corrwg, Garw, Ogwr and Rhondda, the best coals lie below the surface and vertical shafts have been concentrated along the valley bottom—in order to gain depth, as it were. Even so, some shafts are quite deep— that of the Merthyr Vale Colliery (near the tragic village of Aberfan) is about 3,000 feet deep. The concentration of the numerous pit shafts along the valley bottoms, in order to reap the harvest of coal often deep below, resulted in the almost feverish spread of habitation along each valley side. The Rhondda conurbations, for example, are virtually continuous towns straggling in choked-like fashion down the two valleys for distances of up to ten miles, though each portion has its own name —Porth, Pentre, Tylorstown, Treorchy—its own identity and character.

THE EVOLUTION OF THE RIVER SYSTEM

Numerous rivers flow across the coalfield, some to join up with each other before reaching the borders of the coalfield. The river pattern is shown in Figure 75. One outstanding feature of that pattern is the way in which most of the rivers flow *across* the coalfield without too much regard for the underlying geology. Rivers like the Taff, Rhymney and Sirhowy rise on the Millstone Grit, Limestone or Old Red Sandstone, north of the coalfield, then flow southwards across the various divisions of the Coal Measures and thence back over the older Carboniferous rocks and Devonian south of the coalfield before they reach the sea. They thus cut across hard and soft rock formations alike, soft units such as the Lower and Middle Coal Measures, resistant units such as the Pennant Sandstone, Millstone Grit and the Limestone. Moreover, the rivers frequently ignore fault lines, either flowing at some distance away from fractures (for example the Rhondda Fawr) or even crossing them obliquely.

A river which flows in this way across the various geological formations and fractures, without being influenced in any major way by the underlying geological pattern, is said to be *superimposed*, and it certainly looks as if these coalfield rivers have been superimposed on the rocks of the area. In order to explain

FIG 75

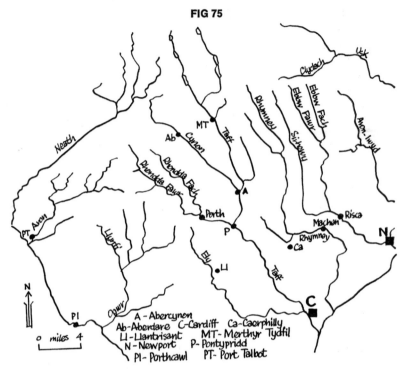

A - Abercynon
Ab-Aberdare   C-Cardiff   Ca-Caerphilly
Ll-Llantrisant   MT- Merthyr Tydfil
N-Newport   P- Pontypridd
Pl- Porthcawl   PT- Port Talbot

such a superimposition, covers of more uniform, younger strata
have been invoked, so that the rivers would first flow across a
homogeneous blanket (Figure 77a). When this cover was eventually pierced, the river would then continue to flow in its same
track even though it was now flowing across the revealed
different rock units below (Figure 77b).

The general opinion has been held that the rivers of South
Wales were *superimposed* on a blanket of Chalk, laid down over
Wales (and much of Britain) during a great marine transgression
in Upper Cretaceous times (about 90 million years ago). That
Wales was once covered by the Chalk Sea is now generally
believed. In fact, it may well have also been previously covered
by Jurassic seas, as thick Jurassic deposits are now known to
occur in the Bristol Channel and thick Lower Jurassic rocks were
encountered in a deep borehole at Mochras, on the coast of
Merionethshire, North Wales. What is in doubt, however, is
whether any of this Jurassic-Cretaceous cover remained over the

FIG 76

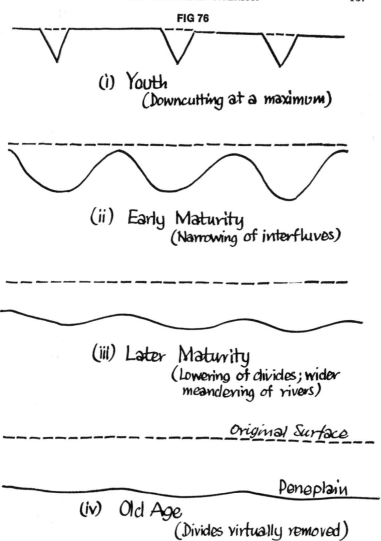

(i) Youth
(Downcutting at a maximum)

(ii) Early Maturity
(Narrowing of interfluves)

(iii) Later Maturity
(Lowering of divides; wider meandering of rivers)

Original Surface

Peneplain

(iv) Old Age
(Divides virtually removed)

Welsh interior when the ancestors of the present-day rivers began to flow. Studies made elsewhere in Britain have shown that earth movements caused folds and faults (with large displacements) to form on several occasions *after* the deposition of the Chalk and these structures would have been actively eroded almost as soon as they formed, thereby helping to remove much

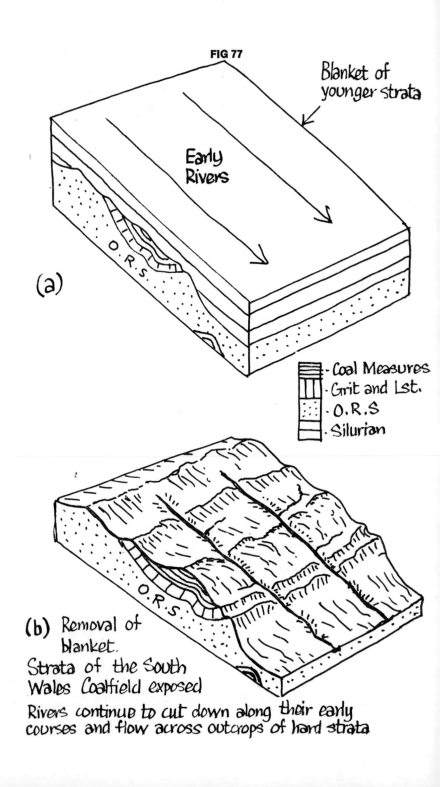

FIG 77

Blanket of younger strata

Early Rivers

O.R.S

(a)

- Coal Measures
- Grit and Lst.
- O.R.S
- Silurian

O.R.S

(b) Removal of blanket.

Strata of the South Wales Coalfield exposed

Rivers continue to cut down along their early courses and flow across outcrops of hard strata of the cover well before the initiation of the modern drainage (probably in Pliocene times, about 5 million years ago).

Moreover, the belief is held by some that a 'stairway' of high-level platforms, ranging today (in South Wales) from over 2,000 feet to 200 feet OD, emerged in succession from a falling Pliocene-early Pleistocene sea level, with longer pauses producing the more extensive plateau surfaces at about 1,800, 1,400, 1,000, 800, 600, 400 and 200 feet OD (the higher the level, the older it is). As these bevelled surfaces emerged from the sea, the early drainage began to form, *superimposed* on the complex geological pattern of Palaeozoic rocks making the newly emergent land surface. With the lowering sea level, more and more land emerged and the rivers flowed further out to the new peripheries—but still superimposed on a complex geological pattern. Eventually, local adaptations, especially by tributaries, caused diversions to the original ancestral stream pattern, but nevertheless the drainage remained essentially a superimposed one. A variation to this theory is that the high-level platforms, now seen on the interfluves between the numerous valleys, were levelled by stream action flowing for long periods of time to static sea levels. The surfaces are thus 'peneplains', this is, relatively level surfaces produced by subaerial—not marine—stream erosion, the result of a long cycle of river erosion (see Figure 76). When each stream cycle reached maturity, uplift of the land (or a fall in sea level) began a new ('rejuvenated') cycle of erosion with the next surface being produced at a lower level. The true explanation may well lie, of course, somewhere between these two alternatives. In other words, the higher surfaces may be peneplains whilst the lower ones (below 800 feet) may be marine (sea-bevelled) platforms.

That some diversions to the original ancestral river pattern (south-south-eastward directed as far as this area was concerned) has occurred is obvious, however, from a glance at the present-day river system. The River Avan, for example, flows across the early pattern in a west-south-westward direction, whilst the River Rhymney makes a marked eastward turn near Caerphilly and then breaches the outer rim of the coalfield at Machen. The lower portion of the Rhondda Fawr, too, turns eastwards prior to its confluence with the Taff at Pontypridd. Wind gaps, now dry, help to reconstruct the early pattern. Two marked cols,

F

at Pen-y-graig and Trebanog, just south of the point where the Rhondda Fawr turns east at Trealaw, mark the 'ghost channels' of the Rhondda rivers that originally flowed southwards as the River Ely, breaching the southern edge of the coalfield near Llantrisant. A right-bank tributary of the Taff (or Aman) developed to such an extent (probably along an intense zone of WNW-ESE joints) that it eventually captured the two Rhondda headwaters of this ancestral Ely. Intense zones of jointing appeared to have also influenced the development of the River Avan, diverting westwards headwaters such as the Corrwg and Pelena and leaving wind gaps such as that between Cymmer and Maesteg.

The eastward diversion of the Rhymney near Caerphilly seems to be related to the presence of a minor syncline within the main coalfield basin. Along this Caerphilly Syncline the highest Upper Coal Measures are preserved in the downfold (Figure 74) and these strata are more shaley and thereby softer than the Pennant sandstones beneath. Those tough sandstones form the strong ENE-WSW ridge of Caerphilly Mountain (or Common) to the south of that town. Figure 74 also shows the parallel ridges formed by the Carboniferous Limestone and the Pennant Sandstones in the vicinity of Caerphilly and Taff's Well.

The present-day pattern of the Taff, Rhymney, Sirhowy and Ebbw Rivers is an interesting one and presents several problems. The most important problem is the N-S portion present in each of these valleys, interrupting the continuity of a more predominant NNW-SSE trend. Such N-S segments are to be seen between Abercynon and Pontypridd (in the case of the Taff), Bargoed and Caerphilly (in the Rhymney Valley), Aberbeeg to Crosskeys (in the Ebbw). One interesting theory (by D. B. Norris) is that after the initiation of the NNW-SSE flowing ancestral drainage (on a south-south-eastward gently tilted surface), subsequent slight warping resulted in N-S tributaries forming on the *left banks* of the parent rivers. These tributaries eventually succeeded in capturing the headwaters of the neighbouring parent stream, thereby producing the present disjointed pattern (see Figure 78 a, b and c).

Long after these modifications to the first drainage pattern, the valley glaciers of the last Ice Age gouged deeply into each valley floor, producing the steep sides. The eroded glacial debris was taken downstream and dumped when the glaciers (or combined

FIG 78

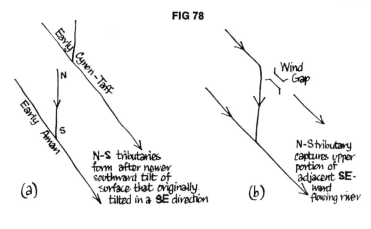

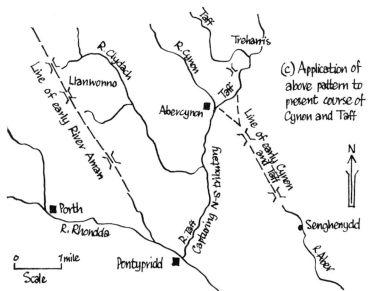

(c) Application of above pattern to present course of Cynon and Taff

ice sheets) melted. At Talbot Green (near Llantrisant) a great glacial mound can be seen. This moraine is crescent-shaped and concave upstream. It extends for 1,500 yards from one flank of the Ely Valley to the other. The thick coarse deposits of this moraine can be seen in a cutting on the Talbot Green bypass, just north-east of the bus station. Another much more moundy moraine can be seen on the golf course at Radyr, five miles north-west of Cardiff. This mound of clayey sand and gravel, over sixty feet thick, was just one portion of the debris dumped by a melting ice front which extended from near Newport westwards to Margam and Swansea Bay. The view northwards from Radyr is a spectacular one. The striking ridge of Garth Wood and Fforest Fawr (formed of tough Old Red Sandstone succeeded by resistant Carboniferous Limestone) is breached by the Taff gorge. New road sections through the gorge reveal several minor folds, especially in the Lower Limestone Shales at the base of the Limestone. Dominating this high ridge is Castell Coch ('Red Castle'—so called because of its red stone), a relatively modern pseudo-castle built by the Marquis of Bute in the style of the castles of Rhineland. It stands on the site of a thirteenth century castle which itself postdated the fortress of Ivor Bach ('Little Ivor'), a short, stocky chieftain who is reputed to have broken into Cardiff Castle and carried off William, Earl of Gloucester, his wife and his son.

The Carboniferous Limestone exposed in the Taff Gorge, here between Tongwynlais and Taff's Well, is itself very reddened— due to staining by red iron oxide. The Triassic cover must have lain at no great distance above this ridge. In fact Sir Aubrey Strahan believed (in 1902) that the present gorge represented an 'exhumed' ancient Triassic valley.

The relationship of geology and topography is excellently demonstrated in this Taff's Well area. The strong ridge of the Limestone and the underlying red sandstones has already been noted. North of the ridge is a major ENE-WSW hollow, drained by tributaries of the Taff and marking the outcrop of the softer shales of the Millstone Grit and of the Lower and Middle Coal Measures. Then comes another strong ridge (Garth Hill on the west and Caerphilly Mountain on the east) again breached by the Taff. This ridge, of resistant Pennant Sandstone, is succeeded northwards by the hollow of the softer highest Coal Measures in the Caerphilly Syncline. In this depression is the modern Nant-garw colliery with its up-to-date processing of many byproducts from coal.

THE RISCA QUARRIES

Interesting exposures of the Carboniferous rocks on the south-eastern border of the South Wales Coalfield occur in the vicinity of the town of Risca, in the Ebbw Valley, six miles north-west of Newport. The coalfield rim is here well marked topographically, presenting a steep scarp to the Vale of Gwent. The scarp is a composite one, its lower slopes being in resistant Devonian sandstones whilst the limestone, Millstone Grit and the Pennant together form the higher portion of the escarpment. The strata dip fairly steeply inwards around this rim of the coalfield, dips of 40 degrees being common. The highest point on the scarp— Twmbarlwm (1,374 feet OD)—is of tough Pennant Sandstone (Figure 80). It is named after a legendary giant and is an Iron Age fortress. The earthworks of this fort are still visible and make the long climb to the hilltop worthwhile. Also rewarding is the

FIG 79

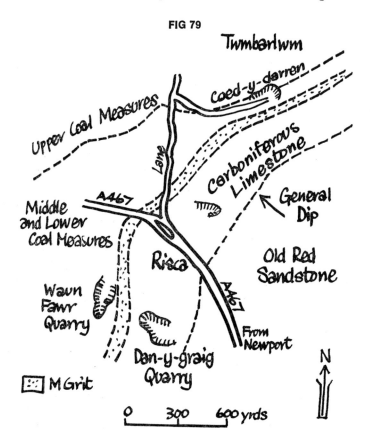

FIG 80

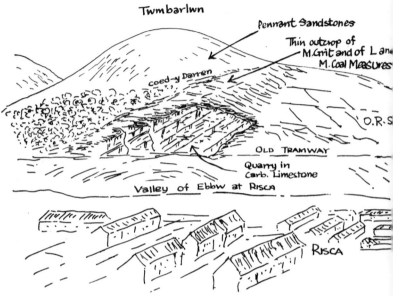

wonderful view from the summit, a view which includes the Somerset coastland together with the Carboniferous Limestone islands of Flat Holm and Steep Holm in the Bristol Channel.

The first of the interesting quarries to visit in the Risca area (Figure 79) are the Dan-y-Graig quarries (ST 234908). These are reached from the main Newport-Crosskeys road (A467) by turning westwards near Risca Station to cross the bridge over the River Ebbw. One can park near the quarries, which expose almost 200 feet of thick-bedded Carboniferous Limestone, medium grey in colour and of *dolomitic* type, that is, the rock contains magnesium carbonate as well as calcium carbonate. Dolomite is needed in the steel industry, particularly for the large Llanwern works near Newport. Another limestone quarry can be seen across the valley. Thin veins of lead have been found in that quarry and in the more recent workings in the Dan-y-Graig quarries. The age of the 'dolomitisation' is puzzling. Some consider that the 'soaking' of the sediments by water containing magnesium carbonate took place when the limestone beds were still plastic (unlithified). Others believe that solutions rich in magnesium carbonate penetrated the hardened limestone after the beds had been formed, even after they had been tilted by folding.

Four hundred yards north-west of these limestone quarries is the Waun Fawr quarry. This can be reached by continuing along the narrow road that passes the limestone quarries, but permission to visit should be asked for in advance. The quarry exposes the lowest beds of the Coal Measures, shales and silty mudstones overlain by a massive quartzitic sandstone (30 feet thick), known locally as 'the Farewell Rock'. The shales and mudstones have been worked for the making of bricks and tiles. Within these softer beds occur a coal (two or three feet thick), known as the Sun Vein, and at least two thin fossiliferous grey mudstone layers containing the brachiopod *Lingula* and scales of the fish *Rhadinichthys*. These fossil bands represent brief marine incursions during the deposition of the otherwise terrestrial Coal Measures. Some measure of retreat and advance of the marine conditions is demonstrated in the exposed quarry face by the marked wedging-out of the strata between the two fossiliferous layers. The presence of these 'marine bands' shows that the 'Farewell Rock' in this area is not of the same age as a formation with the same name in the Vale of Neath or the Swansea Valley. In these more western areas 'the Farewell Rock' sandstone lies *beneath* these marine horizons rather than *above* them.

One further point of interest in the neighbourhood of Waun Fawr quarry is the narrowness of the outcrop of the Millstone Grit. Boreholes drilled recently just south of the quarry showed that only about 75 feet of shales separated the top of the Carboniferous Limestone and a very thick *Gastrioceras subcrenatum* Marine Band (the accepted base of the Coal Measures). Much of the Millstone Grit succession seen further west in areas such as the Neath Valley or Gower must be missing here on this eastern rim of the coalfield, as also is the upper portion of the Carboniferous Limestone.

These striking differences in thickness of the Millstone Grit between the Risca and more western areas are overshadowed by what is to be seen in one other locality in this Ebbw Valley. At Coed-y-Darren, at a height of 800 feet OD on the southern slope of Twmbarlwm, landslip scars reveal that the Lower and

Middle Coal Measures succession is extremely thin—only 250 feet of shales and sandstones, as compared with over 2,500 feet in the more central and south-western areas of the South Wales Coalfield. Even at Wern-ddu, seven miles to the south-west, this succession is already much thicker.

These landslip scars are reached by ascending (on foot) the fairly steep lane leading northwards from the Albert Hotel

FIG 81

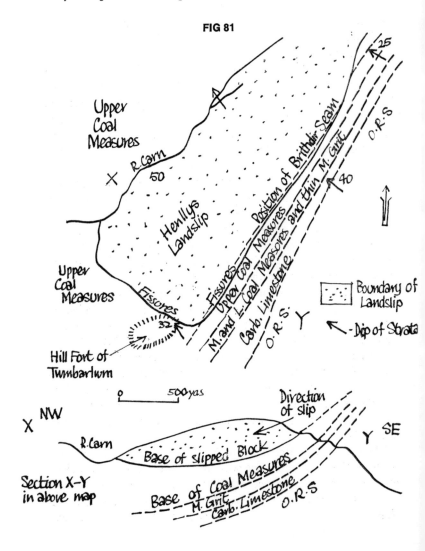

(ST 23479155), on the northern outskirts of Risca. After ascending for about 600 yards, turn off on a track to the right and proceed for another 500 yards to the best exposed landslip scar. The most outstanding formation seen here is a massive pebbly conglomerate, over twenty feet thick and overlain by Pennant-type sandstone in the top of the scar. The base of the conglomerate is believed to be an unconformity. Studies made of these pebbly beds show that the pebbles were brought by rivers flowing from a source area somewhere to the east. Thirty feet below the conglomerate is a thin carbonaceous layer containing coal streaks. This layer is overlain by medium-grey mudstones (even red in patches) containing traces of *Planolites*, a marine worm. This band is believed to be the local representative of the Upper Cwmgorse Marine Band—taken as the agreed upper boundary of the Middle Coal Measures in all the British coalfields.

The broad area of Mynydd Henllys, to the north-east of Twmbarlwm, is a gigantic rotational landslip covering an area of nearly three-quarters of a square mile. Pennant sandstones have slipped north-westwards along a giant curved shear plane developed at about the junction of these sandstones with fireclays and mudstones underlying a prominent coal known as the Brithdir. The foot of the landslip follows the line of the River Carn, a left-bank tributary of the Ebbw. Post-glacial downcutting of the Carn Valley left a large mass of Pennant Measures (predominantly sandstones and dipping north-westwards towards the valley) virtually unsupported and they slid towards the valley. Lubrication between the sandstones and the underlying saturated shales accentuated the slipping (Figure 81).

THE WERN-DDU CLAYPITS

The two claypits at Wern-ddu (ST 168856), one mile south-east of Caerphilly, were excavated (into some 600 feet of the Lower and Middle Coal Measures) for brick-making, using mainly the numerous fireclays (the rootlet beds underlying each coal seam). Twenty-seven separate fireclay bands were exposed in the pits when they were working, some being between five and ten feet thick. Today parts of the pits are overgrown but they are still worth visiting for the partial exposures of this productive portion of the Coal Measures.

The strata dip steeply north-westwards at angles of 30 to 50 degrees. In the smaller, northern claypit, the lowest beds of the Upper Coal Measures are exposed. The northern face of this excavation shows the beds above the Upper Cwmgorse Marine Band with alternations of shales, thin quartzitic sandstones, coal and coaly shales, fireclays and bands of clay-ironstone. Good examples of rhythmic sedimentation occur, each cyclic unit or 'cyclothem' involving a fireclay, coal, a shale roof with plants or 'mussels' and then a sandstone or quartzite (in that upward order). The fireclay represents the muddy slime in which the Coal Measure forests had their roots, whilst the coals were formed of the compressed decayed vegetation (trunks, branches, leaves, etc). The southern slope of the northern claypit partially reveals the Upper Cwmgorse Marine Band which has yielded *Lingula*, *Planolites* and fish scales.

The larger southern claypit still reveals portions of the Middle and Lower Coal Measures. Of particular interest are the sections near the Scouring Brook, showing the split Black Vein coals and crushing along an accompanying fault zone. A waterfall to the south-east reveals thrust faulting at the horizon of the Brass Vein. The same thick coal has a marine roof (the Amman Marine Band) further east along the southern slopes of the larger claypits. A few feet above is the Fork Vein coal with a black roof shale containing non-marine lamellibranchs.

CHAPTER 10

The Vale of Glamorgan

This coastal fringe between Swansea Bay and Cardiff is one of the main agricultural lowlands in Wales. The area, with a gently undulating surface between 150 and 450 feet above sea level, lies south of the coalfield hill-country and has always been a main routeway to the west, in Roman and even earlier times. It is not therefore surprising that the region has a rich history and a glance at a map reveals a large number of ancient British earthworks, burial chambers, ancient castles, churches and priories. The sites of Roman villas occur near Ely and north-west of Llantwit Major, whilst the modern A48 roadway follows a much earlier Roman road. Dunraven Castle (now destroyed) stood on the site of an ancient British castle called Dindryfan and is said to have been the chief palace of the kings of Wales from times as remote as those of Brân ap Llyr and his more renowned son, Caractacus. Ogmore Castle ('Castell Aberogwr' of the Welsh) was the home of William of Londres in the twelfth century, before he went westwards to build Cydweli (Kidwelly) Castle. Margam Abbey, near Port Talbot, was founded in 1147 by one of Henry I's sons. Today, the area is a visitor's paradise, with seaside resorts such as Porthcawl and Barry, wonderful gardens (as at Dyffryn) and a fascinating Folk Museum at St Fagans, near Cardiff.

Topographically, the area reveals the remnants of two major surfaces, both planed by the sea and now lying at near 200 feet and 400 feet above sea level. The 200 foot Platform forms much of the Vale to the south of the main Bridgend-Cardiff road. The cliffs between Dunraven and Rhoose Point are, in the main, 150 feet or over in height, and the rapid erosion of this coastline is shown by the hanging character of several coastal valleys (for example between Dunraven and Nash Point) and by the partial erosion of Iron Age camp encirclements at Dunraven, Nash Point and Sudbrook. Remnants of the 400 foot surface are seen on Cefn Cribwr, near Pyle, and on Cefn Hirgoed. Many parts of the A48 across the vale are further remnants of this higher platform and are the sites of the television masts near St Hilary and Wenvoe. On the southern edge of the coalfield the 200 foot surface gives way rapidly to others at 600 feet or higher. In the west of the vale the rise from lowland to upland is even more sudden. Just inland from the great Margam Steelworks the lowland fringe, below 100 feet OD, rises sharply to over 800 feet on Ergyd Isaf and Graig Fawr. The Port Talbot motorway skirts this scarp on the southern and south-western edge of the great Margam Forest.

DROWNED FORESTS AND BURIED VILLAGES

The most striking features of the Port Talbot-Porthcawl coastline are the extensive ridges of sand known as the 'burrows' or 'warrens', often separating the sea from alluvial flats. These inland flats were once bare but today townships like Aberavon and Port Talbot, and the vast Margam strip-mills, have been built on them. In places sand dunes have encroached on to the inland plateau, as on Merthyr Mawr Warren, east of Porthcawl. Many of these features are the result, directly or indirectly, of the great drowning which occurred at the close of the Ice Age. The ice began to vanish from South Wales about twelve thousand years ago and the return of water to the sea began a prolonged rise in sea level which was to last until the Bronze Age or even later. The total rise in sea level was probably about eighty feet, but this rise was interrupted from time to time by pauses when the land re-established itself and when woodlands clothed the emerged surfaces. Each wooded surface was, however, to be submerged by the next rise of the sea, and as a result several peat layers, with logs and tree stumps, became interbedded with marine and non-marine clays, silts and sands. The uppermost peat layer, with its drowned tree stumps, can be seen today at low tide around the coasts of Swansea Bay. Much of the evidence for these Mesolithic, Neolithic and Bronze to Iron Age drownings has been obtained, however, in the excavations of docks, at Barry in 1895, Cardiff in 1901 and at Port Talbot. Four distinct

peat layers were found at Barry. The lowest, over three feet
thick, was at minus 33 feet OD. Bronze Age bone needles were
found in the top peat, together with the bones and antlers of
red deer. Borings at the lower end of the Swansea Valley proved
peats down to minus 54 feet OD.

With the build up of coastal sand after the last drowning (the
sand possibly first accumulating on the changes of slope which
were once low cliff fringes), large stretches of water were fringed
seawards by the developing coastal sand burrows and warrens
(Figure 82). These 'lagoons' were soon to trap the sediment carried in by streams whose seaward progress was impeded by the
sand dunes, and the mud flats resulted. Who would have thought,
at the beginning of Roman times, that these lagoons would one
day be the site of the Ford works (east of Swansea) or the great
Margam mills! The raising of the sea level after the Ice Age
caused many rivers to deposit their loads of silt and mud, resulting in the silting of many coastal inlets and estuaries. One of
the best examples is the mouth of the River Ely near Penarth,
and of the Rhymney, just east of Cardiff. The new base level of
the River Ely has caused it to shed its load, and it now meanders
in a spectacular manner across the Leckwith Moors and Penarth
Flats.

The coastal fringe east of the Rhymney's mouth is dominated
by the Wentlloog and Caldicot flats. This was also a water area
after the drowning but the drowned land was reclaimed by the
Romans, who built sea walls. Many floods have occurred here
in the past, the most destructive occurring in January of 1607
when ninety-six square miles were flooded to a depth of five feet.

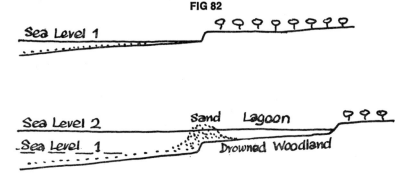

FIG 82

FIG 83

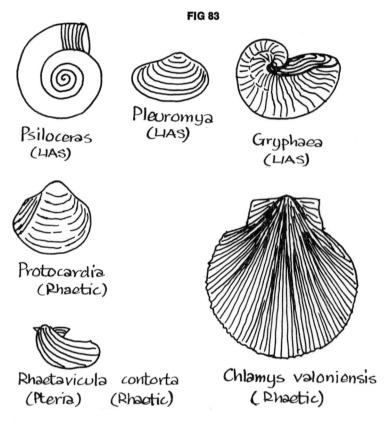

Psiloceras
(LIAS)

Pleuromya
(LIAS)

Gryphaea
(LIAS)

Protocardia
(Rhaetic)

Rhaetavicula contorta
(Pteria)   (Rhaetic)

Chlamys valoniensis
(Rhaetic)

The coastal sand fringes around Swansea Bay have been constantly moving inland, particularly on the Port Talbot-Ogmore side. There is evidence that by Middle Bronze Age times accumulations of sand had reached the limestone hills rising behind Merthyr Mawr Warren. By today the sand has spread over the 200 foot Platform in this neighbourhood, covering a Bronze Age burial mound.

Perhaps the most romantic story concerns the lost village of Kenfig. The inward spread of the sand, even with the prevalent south-west winds, has been sporadic and there have been long periods in historic times when there appeared to be no danger. In Roman times, for example, the first century road—the Via Julia—was built through the Kenfig area. Again a Norman castle was built here. In 1184 the flourishing port of Kenfig was able to

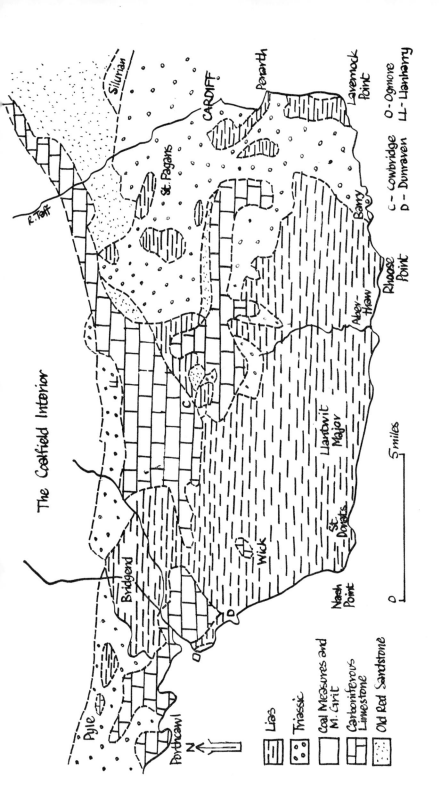

The Coalfield Interior

N

Lias
Triassic
Coal Measures and M. Grit
Carboniferous Limestone
Old Red Sandstone

C – Cowbridge   O – Ogmore
D – Dunraven   LL – Llanhenry

Porthcawl
Pyle
Bridgend
R. Taff
CARDIFF
St. Fagans
Penarth
Lavernock Point
Barry
Rhoose Point
Aber thaw
Wick
Llantwit Major
St. Donats
Nash Point
Silurian

0   5 miles

accommodate two dozen ships. Conditions deteriorated rapidly
however in the thirteenth century, and the church of Maudlam
was moved inland on to higher ground, whilst pasture land was
overwhelmed by sand, as later was the Via Julia. To add to the
interest of this Kenfig area, one must mention the freshwater
pool at Kenfig with its fauna and flora, of importance to students
of biology.

THE GEOLOGY OF THE VALE

The region is important because it is the only area in Wales
where Jurassic strata can be seen *on land* (Figure 84). Moreover,
Upper Triassic rocks also occur, including the interesting Rhaetic
beds, once placed at the base of the Jurassic System but of late
lowered into the underlying Triassic.

At the close of Carboniferous times, earth movements folded
and fractured the Palaeozoic strata and a prolonged phase of
active erosion ensued, lasting for something like 80 million years
—the whole of the Permian period and the lower half of the
Triassic. This erosion removed many thousands of feet of strata
from the crests of anticlines in the Vale of Glamorgan. In the
vicinity of Cowbridge, for example, the whole of the Carboniferous succession—possibly as much as 15,000 feet thick—was
removed prior to the deposition of the Triassic cover. This
tremendous erosion occurred under climatic conditions that can
best be described as sub-arid, with daytime temperatures rivalling
those of the hot deserts today. Rainfall was probably limited to
very short spells in each year, but when it fell it was intense,
and the dry, dusty, rocky wastes were soon converted into sheets
of water with torrents pouring down gullies or 'wadies', situated
on the slopes of higher rocky uplands or 'islands'. These gorge
torrents would carry quite coarse debris, including large boulders, and deposit it as 'fans' around the margins of the lower
plains and basins. The dusts and silts on these plains would be
churned into sticky muds when the rain fell and would accumulate as the deposit known as 'marl'. Some sheets of water would
remain for longer periods, and, under the hot daytime sun, salts
held in solution in the water would be precipitated and would
form lenses and sheets interbedded within the marls. Oxidation
of the deposits would result in reddening of the marls and of the

marly matrix binding the coarser fan deposits (breccias and boulder beds). The salt deposits were frequently in the form known as *gypsum* or *alabaster* (Plaster of Paris is made from gypsum). Masses of gypsum can be seen in the cliffs at Penarth. They occur as the massive, opaque form known as alabaster, and are here stained pink. Fallen masses can be studied on the foreshore, near to the pier car park.

One is bound to ask 'Why was the climate in Triassic times so much hotter and drier than today?' The drier conditions may be partly explained by Britain being perhaps in a rain shadow area behind mountain ranges. There is evidence, however, to suggest that the prevalent winds in Triassic times were from the east or north-east. Moreover, palaeomagnetic studies of Triassic rocks suggest that northern Europe was situated in fairly low latitudes north of the equator, the North Pole being somewhere about Japan! This peculiar position of Britain is part of the wonderful story of 'continental drift'. The continents have moved their relative positions. Some have moved away from one another whilst others have moved towards each other. Australia, for example, has done a waltz around the South Pole on occasions. The South Pole was over Natal, with India tucked in near Madagascar, in late Carboniferous to Permian times. The north-easterly winds over Britain in Triassic times makes sense in fitting in with a position comparable to the present-day Sahara relative to the equator of the early Mesozoic.

A geologist is like a detective, who looks for clues from which he can reconstruct the crime. The geologist looks for clues to help him reconstruct the geography of a past period. Some of these clues have been mentioned already, for example the gypsum deposits and the reddened marls. The finding of coarse boulder beds and angular breccias will indicate to the geologist that they were deposited around the edge of an upland mass, with the finer-grained marls and salt deposits being deposited over the plains and basins, perhaps even in very shallow but extensive waters which connected with the open sea. Plotting the positions and limits of the coarse deposits will help to outline the upland masses. This has been done for the Vale of Glamorgan (and for other areas in Britain, for example the Bristol district). Figure 85 shows the probable outlines of the upland areas in the Vale. Near the uplands, coarse deposits of Triassic breccia

FIG 85

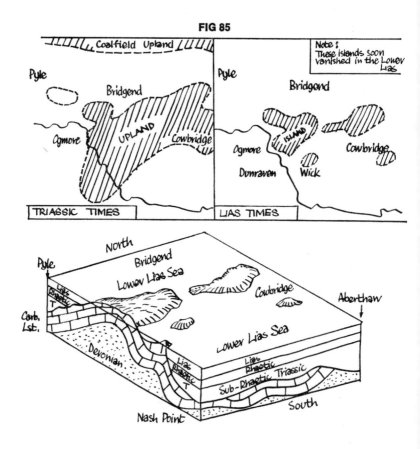

(with angular fragments), conglomerate (more rounded pebbles) and boulder beds have been located. The Triassic deposits are finer grained—marls with occasional salt deposits.

Towards the end of the Triassic period, a rise in sea level resulted in a gradual drowning of the area that is today the Vale of Glamorgan, and the more marine muds and limestones of the Rhaetic were laid down. These strata are best seen at Penarth and nearby Lavernock. Once again, however, the highest upland areas remained above the water, though these residual regions were now less extensive than in the preceding phase. Once again, therefore, finer-grained Rhaetic shales and limestones pass laterally into coarser breccias and conglomerates around the margins of the islands. It is sometimes difficult to differentiate Rhaetic breccias from the earlier Triassic breccias, though in the vicinity of Pyle, and on the northern side of Stormy Down, the coarser Rhaetic marginal deposits are sandy—a formation known as the 'Quarella Sandstone'. Old quarries in it can be seen 100 yards east of the Pyle road sign at the northern approach to the village. The quarries on the northern side of Stormy Down were being worked at the time of writing. The sandy character of the Rhaetic here shows that the hill mass of Stormy Down today is in the same position as an upland mass or island in Triassic times. Parts of Stormy Down must have been above water in the Rhaetic. That it also formed a hill mass, probably even more extensive, in the time of deposition of the red marls of the Triassic is demonstrated by the occurrence of coarser breccias and conglomerates on the eastern flanks of the Down. These coarser beds, resting on Carboniferous Limestone, can be seen just west of the roundabout at the eastern end of the hill (junction of the A48 and A4106). From Stormy Down one has a fine view of the spectacular Pennant scarp overlooking Margam and Port Talbot. Such a marked change of slope must have occurred in the same area in Triassic times with the coalfield mass standing above the level of the Triassic lowland that is now Swansea Bay.

The Rhaetic islands in the Vale of Glamorgan were to be eventually drowned by the Lower Jurassic ('Lias') sea. This sea probably eventually even covered the South Wales Coalfield. Early in the Lias, however, the islands stubbornly resisted drowning, so that there are marginal deposits even of Lower Jurassic age around the edges of the islands. These coarser Liassic strata are of two types—the 'Sutton Stone' and the 'Southerndown Beds'. Both are named after localities near Ogmore and Dunraven. The Sutton Stone is a conglomeratic, white to cream-coloured limestone. It often contains specks of Galena (lead). The usually overlying Southerndown Beds are blue-grey to brown, thick bedded, mottled limestones, which often weather into thinner nodular sheets. The Sutton Stone can pass both laterally and vertically into the Southerndown Beds. Away from the close margins (and crests) of the Liassic islands, the more normal Lias deposits are very distinctive, rapid alternations of thin muddy limestones and grey calcareous shales. The limestone bands are often less than twelve inches thick but are often nodular with irregular 'lumpy' surfaces, even when the bands are persistent. This formation is the typical 'Blue Lias' of the Lower Jurassic. The coasts of the Vale of Glamorgan and of Lyme Regis in Dorset are the best known localities for studying this interesting formation. The position of the Lias islands in the Vale of Glamorgan is shown in Figure 85. Note the diminished size of the Liassic islands compared with the uplands of Triassic times.

Three areas have been chosen for a more detailed examination of the Triassic and Jurassic rocks in the Vale of Glamorgan. The Ogmore-Dunraven coast is selected because it demonstrates the interesting marginal deposits banked against the Dunraven upland island. In contrast, the Penarth-Lavernock coastal section is a classic area for the study of the more normal Triassic-Lias succession, including the Rhaetic. Last, the Llanharry district provides interesting remnants of fissure deposits and even caves of Triassic times. In this district also occurs the only working haematite mine in Wales.

THE OGMORE-DUNRAVEN COAST

This stretch of coastline extends in a NW-SE direction for two and a half miles from the mouth of the River Ogmore to the prominent headland terminating seawards in Trwyn y Witch ('the witches nose') and on which there once stood Dunraven Castle. Numerous ledges and small embayments occur (see Figure 86) with the more prominent bay of Seamouth immediately to the north of the Dunraven headland. The cliffs are low in the north-west of the section but become higher between Black Rocks and Seamouth. The low cliffs are where Triassic rocks rest on Carboniferous Limestone and the higher cliffs are where the Jurassic strata rest on the Carboniferous Limestone. This latter foundation, on which the Mesozoic strata rest unconformably, is formed of light grey Caninia Oolite in the northernmost part of the section with the overlying, coarsely crinoidal, limestones coming on south of the little embayment called Bwlch Kate Antony (Figure 86). These coarser limestones are extremely fossiliferous as an examination of the flat ledges soon reveals.

FIG 86

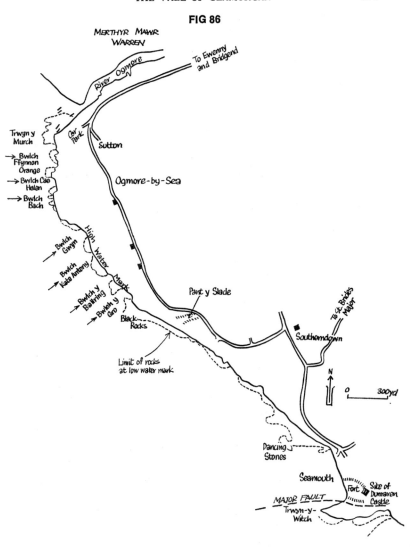

There are three separate exposures of Triassic breccias between the Ogmore's mouth and Black Rocks. In the northernmost, between Bwlch Ffynnon Orange and Bwlch Cae Halen, the breccias and conglomerates are moderately coarse with pebbles up to two feet long. Most of these fragments and pebbles are of Carboniferous Limestone. One boulder is nine feet across. Just

north of the outcrop, several fissures in the underlying Caninia
Oolite are filled with red Triassic sandstone and angular limestone fragments. The central outcrop of breccia, extending for
half a mile to Bwlch Kate Antony, is of finer grain with small,
flattened, angular chips sometimes embedded in grey marl. This
deposit, a great alluvial fan, is up to seventy feet thick. The
southern outcrop, only 60 yards wide, is immediately east of
Bwlch y Ballring and is of much coarser breccias, with many
blocks of Carboniferous Limestone over six feet across and some
even fifteen feet in length. This coarse deposit must have been
caused by 'flash floods' rushing down a steep sided wadi. The
larger blocks were probably the result of the collapse of the
wadi sides. (This reconstruction is the work of T. M. Thomas,
1968.)

All these breccias are thought to be of Triassic age and
deposited around the marginal slopes of a large rocky upland
extending from Ogmore virtually to Cowbridge (Figure 85). They
are considered to be the coarser equivalents of the red Keuper
marls of, for example, the Penarth district. In view of the fact,
however, that around Southerndown and Dunraven the Lias rests
on the Carboniferous Limestone floor, with no apparent Rhaetic
deposits to be seen, it is therefore possible that some of the
Ogmore breccias could be of Rhaetic (or even lowest Lias) age.

The underlying Carboniferous Limestone is particularly fossiliferous between Bwlch Kate Antony and Bwlch y Ballring. Flat
pavements of limestone about fifteen feet above the shore show
wonderful specimens of large corals such as *Caninia*, *Lithostrotion*, *Syringopora*, *Michelinia*, large brachiopods such as
Gigantoproductus and *Chonetes*, gastropods (*Bellerophon* and
Euomphalus) and lots of mats of worm-tubes, the infilled burrows
made by marine worms.

Pant y Slade This is a prominent dry valley running from the
Ogmore-Southerndown road to meet the coast 300 yards east
of Black Rocks (Figure 86). The upper part of the valley shows
the Southerndown Beds, the uppermost of the two marginal
deposits of the Lias. They comprise 80 feet of blue-grey nodular
limestones, occasionally brown and mottled with conglomeratic
layers, towards the base. The pebbles are of chert (silica) and of
Carboniferous Limestone (even containing fossils). At the bottom
of the dry valley is a superb wave-cut platform formed of the

irregular top of the Carboniferous Limestone. Resting on this platform are twenty feet of conglomeratic white to cream-coloured limestone. The fragments are from one inch to several feet across. It is curious that although fragments of Carboniferous Limestone are abundant, there are no fragments of Millstone Grit or Coal Measures. This twenty feet of conglomeratic limestone is the lower marginal deposit of the Jurassic—the Sutton Stone. It is fossiliferous, containing corals (*Isastrea* and *Montlivaltia*), gastropods and lamellibranchs (*Chlamys*, *Lima* and *Cardinia*). The Sutton Stone is also seen in a roadside quarry at Sutton. Here it is massive and not conglomeratic. Specks of galena can be seen.

Seamouth and the Dunraven promontory The inner cliffs of the southern half of Seamouth bay display the normal Blue Lias, alternations of thin limestone with shales. The limestone bands are more predominant in the lower beds of the cliffs. A number of faults can be seen, one being easily recognised by its displacement of a prominent 6 foot limestone. This fault has a southerly downthrow of about 10 feet, bringing the more massive limestone down to near shore level. The most spectacular fault, however, occurs in the south-eastern corner of Seamouth, and trends seawards in an irregular WSW-ENE direction just north of the Dunraven promontory (Figure 86). This fault is a *reverse fault* in that its fault plane slopes towards the side that has gone up. The upthrow side is, in other words, the promontory. As a result of this fault movement, the Carboniferous Limestone floor, with its capping of Sutton Stone and Southerndown Beds, has been brought up (above sea level) on the promontory against the higher, more normal, Lias beds of Seamouth. No wonder then that the tougher Sutton Stone and underlying Carboniferous Limestone have resisted erosion to project as the Trwyn y Witch headland. The pushing up of these more massive formations against the softer Lias shales and thin limestones can be clearly demonstrated in the south-eastern corner of Seamouth, for here can be seen some spectacular minor anticlines and synclines (Figure 87). The limestone bands have been squeezed into lenticles whilst the shales have been polished with the more carbonaceous films almost converted into anthracite coal.

The rapid alternations of thin limestones and shales in the normal Lias of Seamouth yield interesting fossils, including

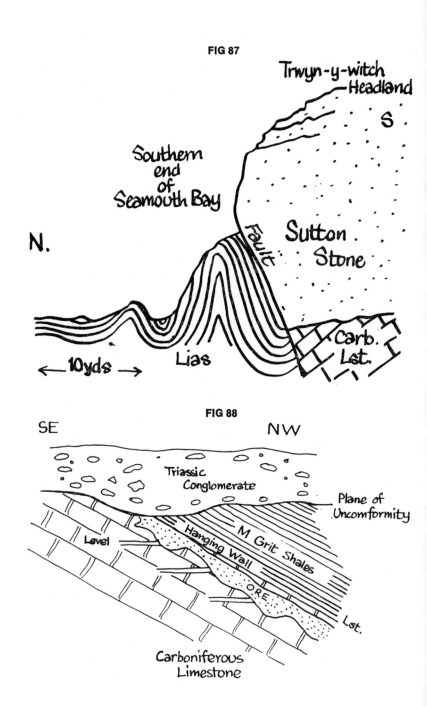

FIG 87

Trwyn-y-witch
Headland

S

Southern
end
of
Seamouth Bay

N.

Fault

Sutton
Stone

Carb.
Lst.

← 10yds →

Lias

FIG 88

SE

NW

Triassic
Conglomerate

Plane of
Uncomformity

Level

M Grit Shales

Hanging Wall

ORE

Lst.

Carboniferous
Limestone

lamellibranchs (*Gryphaea*, *Lima*, *Pinna*, *Pleuromya*), ammonites (*Arietites*) and nautiloids. *Gryphaea* (often called 'the devil's toe nail') is a coiled oyster and many specimens can be collected on the shale ledges near the south-eastern corner of Seamouth. Controversial theories concerning the pattern of evolution of *Gryphaea* communities from more normal oysters (*Ostrea*) have been proposed. Other controversial theories surround the mode of origin of the frequent alternations of limestone and shale in the normal Lias. One theory suggests that the limestone beds were deposited in a primary sense, each independently of the shale horizon above and below it. It has even been suggested that the alternations reflect frequent vertical movements of the sea bed, with the limestone horizons being laid down at times of shallower water. In support of this primary mode of formation is the fact that fossils are frequently eroded at the upper junctions of limestone bands. An opposing idea is that the whole formation was originally deposited as a calcareous mud, but that different rates and intensities of calcite precipitation made certain horizons more calcareous than others. It has been argued here that the fauna of the limestone bands is exactly the same as that in the shales. A third, compromising, idea is that both primary and secondary types of deposition have contributed to this repeated alternation in the Lias. The irregular base of several limestone bands strongly suggests that those heavier layers 'weighed down' on the softer mud layer beneath, producing 'load casting' at the base of the limestone. This would again support a primary origin. Whatever the cause, the result was this very rapid alternation of limestone and shale. At Seamouth, for example, there are 215 limestone bands in 140 feet of Lias succession!

Before leaving this fascinating Seamouth area, a parting examination of the Trwyn y Witch headland is worth making. Looking at the north face of this promontory from the Seamouth sands, one sees the broadly domed cover of massive Sutton Stone, resting discordantly on more folded Carboniferous Limestone beneath. It is worth noting that a cave has been eroded at the unconformable contact of the Sutton Stone with the underlying Carboniferous. Caves are in fact common at this junction between Black Rocks and Seamouth. Standing well back on the sands, one can also try to assess the approximate throw of the

FIG 89

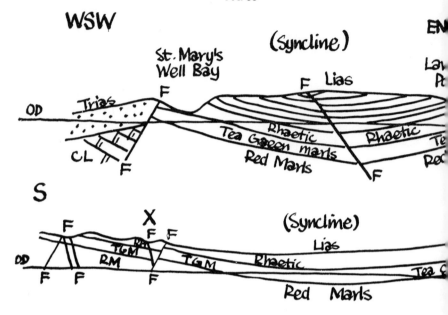

Key: CL - Carboniferous Limestone   RM - R
F - Faults

Note: The two sections continue at the p

important reverse fault in the south-eastern corner of the bay.
The amount of throw must be at least 150 feet.

PENARTH AND LAVERNOCK

The general geology of the area between Cardiff and Barry is
shown in Figure 84. Here the Triassic is more dominantly present,
in its normal Keuper marl type, than at Ogmore and passes up
conformably into the Rhaetic and Lias. The sub-Triassic floor is
again of folded Carboniferous Limestone, as can be seen on Sully
Island and at Barry Island. The Mesozoic rocks are also folded,
into shallow synclines and anticlines, the most obvious downfold
being the Lavernock Syncline. The view from Lavernock Point

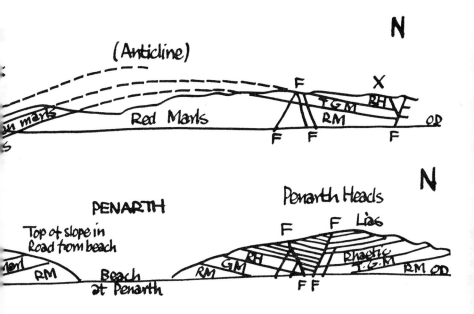

back along the beach towards Penarth clearly shows this broad downflexure, bringing in the greyer Rhaetic and Lias above the red to green Triassic marls. Another syncline occurs in the Penarth Head region, again preserving the Rhaetic and Lias, but here there is also considerable faulting (Figure 89). Many small but nevertheless spectacular faults can be seen along the Penarth-Lavernock section. It is a good section to demonstrate the tilt of the fault planes, direction of downthrow, normal or reverse, combined fault movements producing faulted troughs, etc. Westwards beyond Lavernock Point, in St Mary's Well Bay, there is an important fracture bringing the lowest Triassic sandstones against the black Rhaetic shales. This fault runs inland in a north-westerly direction towards Dinas Powis. These folds and faults were probably the result of the 'Alpine earth movements' which occurred widely during the middle of the Cenozoic era (in the period known as the Miocene).

The full succession to be seen in the Penarth-Sully area is as follows:

LOWER JURASSIC (Lias)		Lias shales (with occasional thin limestones)	-
		Blue Lias—shale and limestone alternations	50 feet
		- - - - - - - - - - - - - -	
T R I A S S I C	Rhaetic	Watchet Beds—grey marls	7 feet
		Langport Beds—pale limestones and shales	2 feet 6 inches
		Cotham Beds—grey to green marls	2 feet 8 inches
		Westbury Beds—blackshales with thin sandstones and limestones	18 feet
		Sully Beds—grey marls	13 feet
		- - - - - - - - - - - - - -	
		Tea Green Marls	30 feet
		Red Marls, passing down into calcareous sandstones	400 feet
		- - - - - - - - unconformity - - - - - - - - - - - -	
		Carboniferous Limestone	-

The junction of the Triassic with the Carboniferous Limestone is best seen on Sully Island. Care must be taken when crossing the narrow causeway to the island—the tide comes in very quickly! The unconformity can be seen at the extreme southeast corner of the island. Horizontal red breccias, with interbedded marls, rest on steeply dipping Carboniferous Limestone.

The cliffs at Penarth, just beyond the covered car park (on the Cardiff side of the sea front) give a good view of the red Keuper marls. These deposits are perhaps best described as fine silts and show thin bands of greenish-grey colour. These green bands increase in frequency upwards and the Red Marls ultimately give way to the 'Tea Green Marls'. The colour change is due to the state of oxidation of the iron in the sediment. The green marls probably accumulated in less arid conditions.

Puzzling, however, are the occasional patches, lenses and even odd spots of green that occur in the main red marls.

Fallen masses of the Tea Green Marls show interesting 'salt pseudomorphs'—casts made by a succeeding marl bed of the cubic impressions made by the solution of rock salt crystals. Of particular interest, of course, are the thin impersistent horizons of gypsum (or alabaster). Many of these masses have fallen on the beach. The salt deposits have always been thought to have been deposited in dried-up inland salt lakes in the Triassic deserts, but recent views suggest that very shallow seas were the scene of the salt precipitation. These seas passed into more open waters at some considerable distance away (perhaps west of the Bristol Channel and in the Irish Sea).

Whilst at Penarth it is worth searching also on the beach for fallen specimens of the overlying Rhaetic formations. Particularly noticeable are the slabs of brownish sandstone displaying well-marked ripple marks. These occur also on the pale limestone slabs of the Langport Beds, together with good 'sun cracks'. The black Westbury shales contain the lamellibranchs *Chlamys valoniensis* and *Rhaetavicula contorta*. *Pleuromya* and *Liostrea* are also common in other Rhaetic rocks. Thin biscuit-coloured limestone layers often abound in the small but distinctive *Protocardia* (Figure 83).

The long walk along the beach from Penarth to Lavernock is worthwhile because the base of the Rhaetic falls virtually to beach level in the centre of the Lavernock Syncline and there are the numerous faults referred to previously. One can save this walk, however, by taking the B4267 road towards Sully and proceeding along the narrow road that leads to Lavernock Point. The descent down a low cliff to the beach level here is usually muddy, so take care.

The junction of the grey Sully Beds with the dark Westbury shales is clearly seen nearby. The uppermost marl bed of the Sully Beds is very waterworn with six-inch deep hollows filled with green marl or black shale. Three bone beds occur in the lowest 3 feet of the Westbury Beds. The lowest is virtually at the base of the beds with the other two at 1 foot 8 inches and 3 feet above that base, respectively. The upper two bone beds are not persistent but the middle one can yield fish remains and quite large reptilian vertebrae.

Above the Rhaetic beds the alternating limestones and shales of the Lias come on at the Lavernock Point. The lowest 20 feet of the Lias yield *Liostrea* but no ammonites. The latter occur, however, in the overlying, more shaley, strata. Crushed specimens of *Psiloceras planorbis* are abundant and near the centre of the syncline (to the west of the Point) there occurs the ribbed *Psiloceras johnstoni*. The highest beds exposed here in the Lias can be examined by climbing a path up the cliff in the very centre of the syncline.

The western limb of this downfold is seen at St Mary's Well Bay. Here also occurs the major fault mentioned previously, the black Westbury shales ending suddenly. Beyond the fracture is the basal Triassic, consisting of calcareous, sandy deposits. Other, more minor, faults can be seen in the Trias. The Carboniferous Limestone cannot be far below.

THE LLANHARRY DISTRICT

Fairly extensive outcrops occur in the vicinity of Llanharry, St Mary Hill and Coychurch, all in the northern part of the Vale of Glamorgan. These areas lay near the northern edge of the large Ogmore-Cowbridge upland of Triassic times and the edge of the coalfield massif was not far away to the north. The coarse Triassic deposits are calcareous breccias, conglomerates and sandstones with bands of calcareous marls. The floor of Carboniferous Limestone is seen in many places and there are numerous examples of joint cavities, open fault gashes and irregular cavities (Triassic caves?) in this floor, all filled with red marly sandstones, fine breccia, inorganic limestones and, in the case of a major cavity (100 feet long by 35 feet deep) in the extreme north-east of the Ruthin quarries (two miles west-south-west of Llanharry), even large angular blocks up to 8 feet long. On an old disused face of the Ruthin quarries, immediately north of the quarry offices, is part of an old Triassic cave. In red marly sandstones of this infilling (above a well defined shaley bedding plane of Carboniferous Limestone) occur an abundance of reptilian bones, usually disseminated through the rock. Occasionally there occur small jaw bones with the teeth attached.

Llanharry Iron-ore Mine Workable occurrences of haematite (iron oxide) in the Vale of Glamorgan have been known since

Roman times. Large-scale working occurred in the second half of the last century and up to 1891 some two and a half million tons of ore were raised in the area between Taff's Well and Llanharry. During World War II Llanharry mine reached an annual output of 200,000 tons in some years. It is the only active iron-ore mine in Wales and when hand-picked averages 55 per cent iron. As in the case of the other occurrences of haematite in the vale, the ore bodies tend to be localised within the top-most beds of the Carboniferous Limestone, under the Millstone Grit shales (Figure 88). At Llanharry these highest limestones become extensively 'dolomitised' (an increasing percentage of magnesium carbonate). Over the mined area the capping of Triassic conglomerate and breccia is 100 to 150 feet thick. West-wards from the existing mine shafts there are five main ore bodies, the major ones being half a mile long (average thickness 100 feet). Sometimes the ore occurs right up to the 'hanging wall' of Millstone Grit shale (see Figure 88) whilst in other places about 30 feet of limestone may intervene. The shafts are 300 feet deep and inclines reach depths of up to 700 feet. A large volume of water is pumped hourly from the mine. Mr T. M. Thomas, who has supplied the above information on the Llanharry area in his book (see reading list on page 200), suggests that some of this water has a deep-seated source, possibly from the large subterranean reservoir within the limestone beneath the coalfield basin. Mr Thomas has suggested that the iron ore could be the result of the chemical replacement of the limestone by iron compounds, the latter being carried downwards either from the Triassic or possibly from the finely disseminated iron pyrite in the Coal Measures. If the latter, then the ore would have been present before the Coal Measures had been removed by pre-Triassic erosion, which followed on the Armorican earth movements.

Glossary

AGGLOMERATE—Coarse, volcanic breccia. Large fragments thrown out by a volcano during explosive activity. Ancient agglomerates show the sites of volcanic activity.

AMMONITE—Coiled-shelled mollusc, in existence from Triassic to Cretaceous times. Rapidly evolving and of wide distribution. Very useful fossils for equating or correlating strata. Most examples less than six inches across but some (in the late Jurassic, for example) up to 4 feet in size.

ANDESITE—Extrusive (volcanic) igneous rock. Named after common occurrences in the Andes. Can be grey-green or purple in colour. Minerals include felspars, hornblende, some augite and mica. Very little quartz.

ANTICLINE—Upfolded or arched structure in strata. Due to compression of the beds. Can be symmetrical or asymmetrical. Erosion of central area (or 'core') of anticline will reveal older strata with younger beds appearing on the flanks or 'limbs'.

ASH—Finer-grained, fragmental material thrown out by volcanic activity. Can be accumulated on either land surfaces or on the beds of seas. In the latter case the ash would be often interbedded with normal sediment and might moreover be even fossiliferous. Composition of the ash related to that of parent magma (molten material). Ashes can therefore be rhyolitic, andesitic, basaltic, etc.

BASALT—Dark, fine-grained lava. Minerals include felspars, augite and olivine. Often cools in columns (as in the Giants Causeway of Ireland) but can also be full of gas holes. Subaerial basalts in the past often associated with red clays or 'boles', the product of tropical weathering.

BATHOLITH—Very large (many miles across) masses of coarse-grained, igneous rock such as granite. Often form roots of mountain chains and often injected during powerful crustal movements. Can be many hundreds of miles across. The space problem may be resolved by having the igneous magma produced 'on the spot' by metamorphic changes of other rocks during intense and prolonged earth movements (the process of 'granitisation').

BIOTITE—Dark brown or black mineral which splits easily into thin scales. Common in granites and in banded metamorphic rocks such as gneiss and schist.

BOSS—Rounded plutonic intrusion, coarsely crystalline. Can be large upward bulge from the upper surface of a batholith. Examples include the Dartmoor Granite.

BOULDER CLAY—Often referred to as 'Drift' or 'ground moraine'. Boulders embedded in a stiff clay. Carried along by ice and dumped by the melting retreating ice. Identification of boulders leads to reconstruction of path of glacier or ice-sheet. Boulder clay surfaces can be wet and marshy. Can also be hummocky with elongate mounds or 'drumlins'.

BRACHIOPOD—Marine, shell-contained animals which have lived in all the geological periods from the Cambrian onwards. Shell of two valves. Each valve symmetrical but shape of the two valves often different. Can be smooth or ribbed, long hinged or short hinged. Often a hole or 'pedicle opening' at top of larger valve. Useful fossils for identification of strata.

BRECCIA—Angular scree deposit, cemented hard. Fragments usually 1 to 10 inches across. Can form along walls of moving faults (fault-breccia). The term 'breccia' is also used to describe angular, fragmented, volcanic rocks, though these should be best termed 'agglomerates'.

CAMBRIAN—Earliest period of the Palaeozoic era. Began 600 million years ago and lasted for 100 million years. The first geological system to yield fossils in any abundance. Main fossils are trilobites.

CARBONIFEROUS—Important system because of its coal seams, limestones and refractories. Some Carboniferous sandstones in Britain contain oil or natural gas. Others are good building stones or aggregates. The Carboniferous period lasted 65 million years. Fossils abundant and include corals, brachiopods, lamellibranchs, crinoids and early vertebrates (fishes and amphibeans). Abundant flora, especially in the Upper Carboniferous.

CENOZOIC ERA—The last 65 million years of time. Divided into a number of periods, the last of which is the Quaternary, noted especially for its Pleistocene Ice Age. The previous periods are generally grouped together as the Tertiary and include the Miocene period, a time of widespread earth movements and mountain building. The Cenozoic (or Cainozoic) era has seen the rapid evolution of the numerous mammal groups, including the horses and the primates (of which Man is the most advanced member).

CHALK—White, pure form of calcium carbonate. Unique formation in the upper part of the Cretaceous System. Forms the 'Downlands' of England. Probably once covered Wales but subsequently eroded. Remnants preserved in Scotland, north-east and south-west Ireland.

CHERT—Form of silica. Like flint, but has a blocky fracture. Common as beds or lines of nodules in the Carboniferous Limestone and near the base of the Millstone Grit in South Wales. Probably first forms as a soft gel, before hardening.

CLEAVAGE—Applied to minerals, the way they split. This is predetermined by the arrangement of the molecules. Mica has a close basal cleavage, calcite a threefold cleavage, quartz has no cleavage. Applied to rocks such as slate, cleavage is a direction of splitting resulting from the compression or folding of weak, incompetent rocks such as mudstones or shales or siltstones. This splitting direction can cut across the bedding at an appreciable angle.

COAL—Formed from peat after long period of change. involving loss of water and gas, bacterial attack and hardening from overlying cover and earth movements. Brown coal and lignite are intermediate forms in the alteration chain from peat to coal. The end processes range from bituminous coal to anthracite with further loss of volatiles and increase of carbon.

G

CONGLOMERATE—Hardened rock made up of rounded pebbles set in a finer-grained matrix (often siliceous). May have originally been a beach shingle or a river gravel. Fossil boulder clays are known as 'tillite'.

CORALS—Living today in warm waters but also abundant in the past. Secreted stony skeletal structures, either as single structures or as massive colonies. Important as limestone formers and as pointers to ancient environments. Common in the Silurian and Carboniferous limestones of Wales.

CRETACEOUS—Last of the Mesozoic periods. Lasted 70 million years. Areas of deposition restricted at first in Britain but later expanded to cover many areas, including Wales. Best known from areas such as the Weald of south-east England. The ammonites and the dinosaurs died out before the close of this period.

CRINOIDS—'Sea lilies', marine animals growing from the sea floor by long stalks at the base of a cup (with arms). Some forms were free-swimming. Crinoidal limestones are those made up largely of broken crinoid debris, especially segments (columnals) of the stalks.

DEVONIAN—Named after Devon. Devonian rocks of Britain (with exception of south-west England) mainly of non-marine origin and deposited in fluviatile, lacustrine and estuarine environments. Alternatively known as the Old Red Sandstone. Devonian rocks of Devon and Cornwall were deposited in a sea that stretched into France and Germany.

DIORITE—Plutonic igneous rock, usually medium to dark grey in colour. Constituent minerals are felspar, hornblende and augite with a little mica. Some types contain quartz.

DIP—The tilt of beds. Angle of dip or tilt measured in degrees from the horizontal. Vertical beds dip at 90 degrees. The direction of 'strike' of the beds is at right angles to the direction of maximum tilt.

DOLERITE—Dark-coloured igneous rock commonly occurring as dykes or sills. Constituent minerals are felspar, augite and olivine. Often have a dark-greenish tinge when wet—hence the old-fashioned term 'greenstone'.

DOLOMITIC LIMESTONE—Fawnish-weathering limestone rich in magnesium carbonate as well as calcium carbonate. Dark grey when fresh. Used in smelting of steel. Common in the Carboniferous Limestone of the south-east crop of the South Wales Coalfield.

DRIFT—Alternative name for boulder clay and other glacial deposits. Geological maps showing distribution of glacial deposits generally known as Drift maps.

DYKE—Relatively narrow igneous intrusion which usually cuts through beds at a high angle. May often weather in a more resistant manner than the softer, invaded sedimentary rocks, and thereby form a wall ('dyke').

EOCENE—Second period of the Cenozoic era. Lasted for 15 million years. Time of intense volcanic and intrusive activity in north-west Britain. Include basalt sheets of the Giants Causeway in Northern Ireland. Sediments of the London and Hampshire basins deposited at this time.

ESCARPMENT—Steep topographical rise, extending for some length. Usually caused by resistant formation, often gently dipping.

EXTRUSIVE ROCKS—Poured out, thrown out or blown out from volcanic vents or fissures. Include lavas, ashes, tuffs and agglomerates.

FAULT—Fracture, movement along which has dislocated the continuity of the strata. Side which has gone down (relatively) referred to as 'downthrow side'. Slope of fault plane measured in degrees from horizontal. Slip along faults may have been more lateral than vertical. Such faults called 'tear' or 'wrench' faults. Faults may combine to either drop an intermediate portion ('trough' or 'graben') or lift that central block ('horst').

FELSITE—Igneous rock, reddish coloured with quartz and felspar. Intrusive.

FELSPAR—Aliminium silicate mineral. Common rock-forming mineral in igneous rocks. Many kinds of felspar. Orthoclase Felspar often pink in colour. Weathers to kaolinite or 'china clay', used in pottery industry.

FIRECLAY—Underclay or 'seat-earth'. The slime or soil in which the Coal Measure forests had their roots. Always underlie coal seams. Sandy soils formed tougher 'ganister'. Full of rootlets. Used in manufacture of bricks or tiles for fireplaces, etc.

FJORD—Over-deepened valley by glacial action. Such valleys have steep sides and often a lip at their mouths. May be drowned later by rises of sea level. Fine examples today in north-west Scotland and in Norway.

FLAGSTONE—Sandy rock which splits into thin beds about an inch or so in thickness. Used for paving flags and walling. Common in the Coal Measures of Britain.

FOSSIL—Any remains of a once living animal or plant. May be bones, casts or moulds, impressions, petrified wood, etc. Occur usually on bedding planes, but fossilised tree trunks, etc may penetrate through several beds. Help to reconstruct conditions in the geological past.

GABBRO—Dark, coarse-grained, igneous rock, usually rich in augite and olivine. Usually large intrusions.

GALENA—Lead. Steel-grey colour, black streak. Occurs as veins, often of quartz, and often with zinc-blende and pyrite. Usually of hydro-thermal origin, associated with igneous phases. Occurs particularly in the Ordovician and Silurian rocks of Central Wales.

GASTROPOD—Coiled mollusc, coiled in an upward spiral rather than in a vertical plane (ammonites). Common in periods like the Silurian, Jurassic and in the Cenozoic rocks. Abundant today.

GNEISS—Coarsely banded, metamorphic rock. Lighter layers usually of felspar or quartz, darker layers of hornblende or other dark minerals. Can be altered product of plutonic igneous rocks or coarse sedimentary rocks.

GONIATITE—Plane-coiled mollusca, abundant in the Devonian, Carboniferous and Permian periods but then died out to be replaced by the Ammonites. Useful fossils in zoning the Devonian (marine) and the Millstone Grit.

GRANITE—Coarse, crystalline, light-coloured igneous rock. Rich in quartz, felspar and mica. Very hard stone. Used as ornamental stone and building stone (City of Aberdeen). Can form from further crustal changes in metamorphic rocks. Granite intrusions usually large and often associated with intense earth movements.

GRAPTOLITE—Fossils of doubtful affinity. Fragile, small and slender, branches resembling the quill of a pen. Vary in branch numbers and in form of individual cups of each branch. One-branched forms (Monograptids) occur in the Silurian rocks. Became extinct, except for related forms, by Lower Devonian. Very useful fossils in identifying horizons in the Ordovician and Silurian systems. Rapidly evolving and of wide distribution. Marine animals, floating on sea surface and falling into bottom muds on death.

GREYWACKE—Gritty rock in which the grains are embedded in a muddy matrix. Usually deposited by fast-moving, turbidity currents, with slumping of sediments down submarine slopes. Common in the Lower Palaeozoic rocks of Britain.

GRIT—Sandstone in which the grains are angular rather than rounded.

HEAD—Angular, scree-like deposit, produced during thaw conditions in the Ice Age. Common around the marginal areas of glaciated Britain.

HOLOCENE—The most recent division of geological time and of the Quaternary period. A time of rises in sea level in Britain and of the dominance of man.

HYPERBYSSAL—Term used for those igneous rocks that form the smaller or thinner intrusions, such as sills, dykes, small laccoliths, etc. Include Dolerite and Quartz-felspar Porphyry.

IGNEOUS ROCKS—Formed from once molten material (magma). Degree of coarseness of crystallisation depends mainly on depth from surface of cooling position of the magma. Therefore deep plutonic masses are coarsely crystalline, whereas surface lavas are fine-grained.

INLIER—Area of older rocks surrounded by younger strata. An example would be the core of an eroded anticline or the bottom of a depression in the land surface.

INTERGLACIAL—Milder phase within a major Ice Age. During the last (Pleistocene) Ice Age, there were at least three interglacials and on one of these occasions the climate in Britain was warmer than today.

INTRUSION—Igneous mass pushed through sedimentary layers. May be large masses which solidified deep below the surface or smaller, narrower sheets or pipes which cooled and crystallised nearer the surface. Subsequent erosion can expose these intrusions, when they usually form outstanding topographical features because of their hardness and resistance to erosion.

JOINT—Fracture in rocks. Differs from a fault in that virtually no slip has occurred along a joint. A joint may of course become a fault on movement being initiated. Can occur in igneous and sedimentary rocks. Usually form sets. Possible to distinguish major (master) joints from less important joints. Erosion tends to be accentuated along joints. In limestone districts water drains underground down joint systems, forming caves and potholes. Joints are useful in quarrying.

JURASSIC—Important period of the Mesozoic era. Time of dominance of the Ammonites and the huge dinosaur reptiles. Jurassic seas covered large areas of Britain at times but were never deep. Tropical climate and vegetation on the land areas.

LAMELLIBRANCH—Bivalved mollusca. Both valves usually of same shape, but each valve tends to be asymmetrical. Teeth and sockets in each valve. Include modern cockles and mussels, razor shells, etc. Important in periods like the Silurian, Carboniferous, Jurassic and Cretaceous. Also very abundant in the Cenozoic rocks, as they are today.

LIAS—Derived from the word 'layers'. Well-layered alternations of lime-stone and shale in South Wales. The Lower Jurassic division. Lias rocks in Britain include important bedded iron-ores, worked on a large scale.

LIMB—Flank of a fold. In the case of an anticline the limbs dip outwards from the axis of the upfold. Vice versa for a syncline.

LIMESTONE—Rock rich in calcium carbonate and also, on occasions, magnesium carbonate (when it becomes a Dolomite). Usually formed in shallow, clear, warm water. Usually very fossiliferous with corals, brachiopods, crinoids, etc. Quarried for lime, industrial flux, building stone, etc.

MARINE BAND—Thin layer, with marine fossils, interbedded in non-marine successions. Indicate a brief incursion by the sea into a non-marine depositional environment. Important horizons in the Millstone Grit and Coal Measures.

MARINE PLATFORM—Flat uplifted surface, bevelled by a higher sea level and then raised. Common at heights of 200, 400 and 600 feet around the coasts of southern and western Britain.

MARL—Fine-grained, muddy sediment containing some lime. Red marls common in the 'Old Red Sandstone' and in the Permian-Triassic systems. Gives rich soils, good for agriculture.

MESOZOIC ERA—The third era of geological time. Lasted from 225 until 65 million years ago. The time of ammonites and reptiles. In Britain, virtually free of igneous activity. 'Middle life'.

METAMORPHIC ROCKS—Rocks which were previously sedimentary or igneous but which were subsequently altered by heat or pressure (or both), as a result of intense earth movements or igneous intrusion. Often banded or very micaceous. Marble is altered (baked) limestone. Usually hard rocks and form rough, rugged country, often with poor soil cover.

MICA—Extremely fissile mineral, scales easily. Dark form is biotite, transparent variety is Muscovite Mica. Mica common in granite and metamorphic schists.

MIOCENE—Time of intense earth movements in many parts of the world. Miocene rocks missing in Britain.

MONOCLINE—S-fold with a vertical (or nearly vertical) middle limb. Examples include the Isle of Wight fold and the Lulworth fold, Dorset.

MORAINE—Mound of glacial debris deposited by melting ice. May be a terminal moraine (at the furthest point of advance of a glacier) or a recessional moraine (at pause points in the retreat of a glacier) or even a lateral moraine (debris deposited along the sides of a glacier).

NAPPE—Large-scale flat overfold in which the two limbs are virtually horizontal. It follows that on the underside limb the succession will be inverted. Nappes are common in the Alps and in the Precambrian areas of northern Scotland.

NEOGENE—Name given to the second half of the Tertiary, that is, to the Miocene and Pliocene divisions.

NORMAL FAULT—Fault in which the plane of the fracture slopes towards the downthrow side. The opposite case is that of a Reverse Fault.

OOLITE—Limestone in which the internal structure consists of small, concentric bodies of lime, resembling the roe of a fish. Portland

Stone is a good Upper Jurassic oolitic limestone.

ORDOVICIAN—Important period in the Lower Palaeozoic. Time of appreciable igneous activity. The mountains of Snowdonia and the Lake District are carved out of resistant Ordovician volcanic rocks. Graptolites and trilobites the chief fossils. Named after a North Wales ancient tribe—the Ordovices.

OUTLIER—Area of younger rocks surrounded by older strata.

OVERFOLD—Fold in which one limb has tilted beyond the vertical.

PALAEOZOIC—'Ancient Life'. The era ranging from the Cambrian to the Permian. Lasted 375 million years and saw the evolution of many invertebrate and vertebrate groups of animals—trilobites, graptolites, fishes, corals, amphibeans, etc. Generally divided into a lower and upper portion.

PERIOD—Division of geological time. Main eras are divided into a number of periods.

PERMIAN—Last period of the Palaeozoic. Named after a province in the Russian Urals. Time of extinction of many Palaeozoic animal families, including the trilobites and tabulate corals. Permian rocks are absent in Wales.

PILLOW LAVA—Submarine volcanic extrusion in which the lava congeals in heaped masses resembling pillows. The base of a pillow tends to be flat or bulged upwards. Common in the Welsh Ordovician volcanics.

PLEISTOCENE—Major portion of Quaternary time. The time of the Great Ice Age. Time of rapid evolution of man and his flint cultures.

PLIOCENE—Last of the Tertiary periods and preceded the Pleistocene. Pliocene deposits restricted in Britain and occur mainly in East Anglia. Time of appreciable changes in sea level, accounting for the flat platforms around the British coasts, if not even also for higher planed surfaces across the British hills and mountains.

PLUTONIC—Deep-seated igneous masses formed of coarse rocks such as granite and gabbro.

PORPHYRITIC—Texture in igneous rocks (usually in sills and dykes). Larger, well-formed crystals are set in a fine-grained groundmass, showing that there were two distinct phases of cooling and crystalisation.

PRECAMBRIAN—Name given to the rocks of the extremely long time before the Palaeozoic. Precambrian rocks can therefore be any age between 600 million years old and the origin of the earth (almost 5,000 million years ago). Rocks as old as 2,600 million years are known to occur in north-west Scotland. Fossils are rare in the Precambrian, though this may be due to the fact that life was mostly soft-bodied. Large areas of Precambrian are formed of metamorphic gneisses and schists, as in northern Scotland.

QUARTZ—Mineral formed of silica. Abundant as grains in grits, sandstones and conglomerates. Abundant also in granites and in rhyolites. Common lighter bands in gneisses. Also occurs widely as veins, especially in areas of igneous activity. Quartz veins are frequently sources for lead, zinc, tin, copper, gold, etc.

QUARTZITE—May be a sedimentary rock, when the quartz grains are set in a quartzose matrix. May also be a metamorphic rock with a mozaic of quartz crystals. Quartzites are common in the Millstone Grit of Wales, where they are quarried for refractories.

RAISED BEACH—Layers of cemented shingle, often with sea shells, now found well above modern sea level. Examples at about 30 feet above sea level in Gower and on the Isle of Portland. Many point to changes of sea level during the last Ice Age.

RHYOLITE—Light-coloured volcanic lava, rich in quartz and felspar. Extremely hard rock and occurs commonly in the Ordovician igneous areas such as Snowdonia.

RIA—Drowned inlet extending a considerable way inland. A drowned valley system. Examples include the Solva ria and the mouth of the Dart in Devon.

SANDSTONE—Consolidated and cemented sand, in which the grains are fairly well-rounded. One of the most common of the sedimentary rocks. Matrix may be siliceous, calcareous or ferruginous. Sandstones with well-rounded grains and a loose, friable cement are good porous horizons and can store water, oil and gas.

SCHIST—Glossy, micaceous, metamorphic rock. Alteration of shales, slates, mudstones, fine-grained sandstones and igneous rocks such as dolerites, basalts, etc.

SEDIMENTARY ROCKS—Hardened, cemented products of once soft, friable sands, muds, gravels, screes, etc. Also include rocks of organic origin (coals and certain limestones) and of chemical origin (salt deposits, chemically formed limestones, etc). Sedimentary rocks are characteristically stratified or bedded.

SHALE—Hardened muddy rock which splits into thin leaves parallel to the bedding.

SILL—Intrusive igneous sheet parallel to the stratification of the sedimentary layers into which it is intruded. Distinguished from an interbedded lava flow by having a zone of baking *above* as well as below the sill. Sills often of dolerite but can be of felsite or of porphyry.

SILTSTONE—Intermediate in grain between a shale (or mudstone) and sandstone. Hardened product of silts.

SILURIAN—Palaeozoic period that ended 395 million years ago. Chief fossils include trilobites, corals, brachiopods and single-branched graptolites. Primitive plants and first fishes appear in this period. Best known Silurian rocks occur in Shropshire and include the limestones of Wenlock Edge.

SLICKENSIDES—Grooves and scratches produced on each rock wall when slip occurs along a fault. Often also polished by the movement. Slickenside will be parallel to the direction of movement. Therefore horizontal slickensides indicate tear faulting.

STRATIFICATION—Layering or bedding in sedimentary or extrusive igneous rocks. May be result of colour changes or changes in rock type or may result from brief pauses in deposition. Beds may vary in thickness. Beds over 6 feet thick can be termed 'massive'.

STRATUM—Bed or layer. Plural: 'strata'.

SYNCLINE—Downfold. Limbs dip inwards towards axis of downfold.

SYSTEM—Term denoting the thickness of strata deposited during a period. Therefore the Carboniferous Limestone, Millstone Grit and Coal Measures together make up the Carboniferous System. Lower Carboniferous and Upper Carboniferous *time* make up the Carboniferous Period.

TEAR FAULT—Fracture along which horizontal (or nearly horizontal) slip has taken place. Many of the world's greatest fractures are tear faults, as for example the San Andreas Fault of California or the Alpine Fault of New Zealand. The relative slip along the Great Glen Fault of Scotland (under Loch Ness) was as much as 68 miles.

TERTIARY—The combined Palaeocene to Pliocene periods of time. Old term which originated when the geological succession was subdivided into Primary (Palaeozoic and older), Secondary (Mesozoic), Tertiary and Quaternary.

THRUST—Low-angled fault where the upper side has been pushed over the underside. Often associated with overfolds and nappes. May carry rocks for a considerable distance. An example is the Moine Thrust of north-west Scotland.

TRIASSIC—First period of the Mesozoic. Time of hot arid conditions in Britain. Name originates from the German succession which can be subdivided into three parts: the Bunter, Mushchelkalk and Keuper (in that upward order). In Britain the marine middle portion is missing and is replaced by the upper part of the Bunter Sandstone and the lower portion of the Keuper Marl. In South Wales only the Upper Keuper is present of the Triassic succession.

TRILOBITE—Marine animals with a hardened outer cover. Invertebrates. Body divided into head, chest and tail (pygidium). Chest and tail segmented with a central axis. Mode of life variable. Some floated, others were sessile, some burrowed. Useful for zoning the Cambrian System. Largest specimens ,about 3 feet long, found in Cambrian of Pembrokeshire.

TUFF—Fine volcanic 'soot'. Often settled with marine muds and can yield fossils.

TURBIDITY CURRENTS—Fast-moving currents carrying clouds of murky sediment down submarine slopes and along the bottoms of submarine canyons. Characteristic of the Continental Slope, that is beyond the edge of the Continental Shelf.

UNCONFORMITY—Break or discordance in the rock succession of a region. May be indicated by a marked discordance in dip or by the proved absence of a part of the regional sequence. There may even be many systems missing, as for example when Triassic rocks rest on Ordovician (as in north-east Wales). In parts of Pembrokeshire, Silurian rocks rest on Precambrian.

VOLCANIC ROCKS—Lavas, ashes, tuffs and agglomerates. The products of volcanic action.

WELDED TUFF ('IGNIMBRITE')—Volcanic rocks formed under great heat, when the volcanic fragments (or 'shards') are welded together, a process like soldering. Indicates subaerial vulcanicity and occurs particularly in the British Ordovician.

Bibliography and References

GENERAL GEOLOGICAL INTEREST

Ager, D. V. *Introducing Geology* (Faber & Faber)
Anderson, J. G. C. and Owen, T. R. *Structure of the British Isles* (Pergamon)
Bennison, G. M. and Wright, A. E. *The Geological History of the British Isles* (Arnold)
Kirkaldy, J. F. *Minerals and Rocks* (Blandford Press)
Rhodes, F. H. T. *The Evolution of Life* (Pelican)
Wells, A. K. and Kirkaldy, J. F. *Outline of Historical Geology* (Murby)
Fossils : A little guide in colour (Paul Hamlyn)
Mesozoic Fossils (British Museum, Natural History)
Minerals : A little guide in colour (Paul Hamlyn)
Palaeozoic Fossils (British Museum, Natural History)

SOUTH WALES GEOLOGY

Anderson, J. G. C. *Geology around the University Towns. The Cardiff District.* Geologists Assoc. Guide No 16 (Benham & Co). (Also Owen, T. R. and Rhodes, F. H. T. for *The Swansea area*, Guide No 17)
Basset, D. A. 'A source-book of geological, geomorphological and soil maps for Wales and the Welsh Borders (1800-1966)', Cardiff
Bassett, D. A. and Bassett, M. G. (editors) *Geological Excursions in South Wales and the Forest of Dean* (Geologists Assoc, South Wales Group)
Brown, E. H. *The Relief and Drainage of Wales* (Univ Wales Press)
George, T. N. *British Regional Geology, South Wales* (HMSO)

199

North, F. J. *The River Scenery at the head of the Vale of Neath*
4th ed (National Museum of Wales)

Squirrell, H. C. and Downing, R. A. *Geology of the South Wales
Coalfield, Part 1. The Country around Newport (Mon)*, 3rd ed
Mem Geol Surv UK

Thomas, T. M. *The Mineral Wealth of Wales and its Exploitation*
(Oliver and Boyd)

Woodland, A. W. and Evans, W. B. *Geology of the South Wales
Coalfield, Part 4. The Country around Pontypridd and Maesteg*,
3rd ed Mem Geol Surv UK

MAPS

Ord Surv. One inch maps: 138 (Fishguard), 139 (Cardigan), 140
(Llandovery), 141 (Brecon), 128 (Montgomery), 151 (Pembroke),
152 (Carmarthen and Tenby), 153 (Swansea), 154 (Cardiff), 155
(Bristol and Newport)

Geol Surv. One inch maps: 226-7 (Milford), 228 (Haverfordwest),
229 (Carmarthen), 230 (Ammanford), 231 (Merthyr Tydfil), 232
(Abergavenny), 244 (Linney Head), 245 (Pembroke), 246 (Worms
Head), 247 (Swansea), 248 (Pontypridd), 249 (Newport), 261-2
(Bridgend), 263 (Cardiff)

PAPERS OF SPECIAL INTEREST

Allen, J. R. L. 1963. 'Depositional features of Dittonian rocks:
Pembrokeshire compared with the Welsh Borderland', *Geol
Mag* 100, 385-400

Baker, J. W. 1971. 'The Proterozoic history of Southern Britain,
Proc Geol Ass, 82, 249-66

Bassett, D. A. 1968. 'Directory of British geology. 1 A provisional
annotated bibliography and index of geological excursion
guides and reports for areas in Britain, 2 Wales and the Welsh
Borders', *Welsh Geological Quarterly*, 3 no 1 (1967), 3-23

Bluck, B. J. and Kelling, G. 1963. 'Channels from the Upper
Carboniferous Coal Measures of S. Wales', *Sedimentology*,
2. 29-53

Coase, A. C. 1967. 'Some preliminary observations on the geomorphology of the Dan-yr-Ogof Caves'. *Proc Brit Spel Ass*, 5.
53-67

Cox, A. H., Green, J. F. N., Jones, O. T. and Pringle, J. 1930. 'The geology of the St David's district, Pembrokeshire', *Proc Geol Ass*, 41. 241-73

Garwood, E. J. and Goodyear, E. 1919. 'On the geology of the Old Radnor district', *Q Jl Geol Soc Lond*, 74. 1-30

George T. N. 1940. 'The structure of Gower', *Q Jl Geol Soc Lond*, 96. 131-98

George, T. N. 1961. 'The Welsh Landscape', *Sci Progr Lond*, 49. 242-64

Jones, D. G. 1969. 'The Namurian succession between Tenby and Waterwynch, Pembrokeshire', *Geol J*, 6. 267-72

Jones, O. T. and Pugh, W. J. 1949. 'An early Ordovician shoreline in Radnorshire, near Builth Wells', *Q Jl Geol Soc Lond*, 105. 65-99

Jones, R. O. 1939. 'The evolution of the Neath-Tawe drainage', *Proc Geol Ass*, 50. 530-66

Kellaway, G. A. 1971. 'Glaciation and the Stones of Stonehenge'. *Nature*, 232, no 5314. 30-5

Kelling, G. 1968. 'Patterns of sedimentation in the Rhondda Beds of S Wales', *Bull Am Ass Petrol Geol*, 52. 2369-86

North, F. J. 1955. 'The Geological History of Brecknock', *Brycheiniog*, 1. 9-77

Owen, T. R. 1954. 'The structure of the Neath Disturbance between Bryniau Gleision and Glynneath, S Wales', *Q Jl Geol Soc Lond*, 109. 333-65

Owen, T. R. Rhodes, F. H. T., Jones, D. G. and Kelling, G. 1965. 'Summer (1964) Field Meeting in South Wales', *Proc Geol Ass*, 76. 463-96

Owen, T. R., Bloxam, T. W., Jones, D. G., Walmsley, V. G. and Williams, B. P. J. 1971. 'Summer (1968) Field Meeting in Pembrokeshire, S Wales', *Proc Geol Ass*, 82. 17-60

Owen, T. R., and others, 1971. 'Excursions A2 and B2. South Wales', 6th Int Congr on Carb Strat Geol, Sheffield

Simpson, B. 1955. 'Field Meeting in South Wales (1951)', *Proc Geol Ass*, 65. 328-37

Thomas, G. E. and Thomas, T. M. 1956. 'The volcanic rocks of the area between Fishguard and Strumble Head, Pembrokeshire', *Q Jl Geol Soc Lond*, 112. 291-314

Thomas H. H. 1923. 'The Source of the Stones of Stonehenge', *Antiquaries*, 3. 239-61

Thomas, H. H., and Cox, A. H. 1924. 'The volcanic series of Trefgarn, Roch and Ambleston', *Q Jl Geol Soc Lond*, 80. 520-48

Thomas, T. M. 1959. 'The geomorphology of Brecknock', *Brycheiniog*, 5. 55-156

Thomas, T. M. 1968. 'The Triassic rocks of the west-central section of the Vale of Glamorgan with particular reference to the "boulder" breccias at Ogmore-by-Sea', *Proc Geol Ass*, 79. 429-39

Wobber, F. J. 1965. 'Sedimentology of the Lias (lower Jurassic) of S Wales', *J Sedim Petrol*, 35. 683-703

Waltham, A. C. 1971. 'A note on the structure and succession at Abereiddy Bay, Pembrokeshire', *Geol Mag*, 108. 49-52

Ziegler, A. M., McKerrow, W. S., Burne, R. V. and Baker, P. E. 1969. 'The age and environmental setting of the Skomer Volcanic Group, Pembrokeshire', *Proc Geol Ass*, 80. 409-39

Appendix

Acknowledgements

I wish to express my sincere thanks to my colleagues in the Department of Geology, University College of Swansea, for many stimulating discussions on the geology of South Wales. I would like particularly to single out Dr Gilbert Kelling in this respect. Miss Mair Davies typed the chapters and Mrs Alvis Smith photocopied many of my diagrams. Mr S. Osborn photographed my other diagrams. To them I extend my thanks also. I am particularly grateful to Mr J. U. Edwards for specially drawing Figures 21, 22 and 27.

Dr Prys Morgan has kindly supplied me with the information on places of historic interest. To him I give my grateful thanks.

For the inspiration of this book I am grateful to the hundreds of people who have attended my extra-mural classes. Their company, both in the lecture room and on many geological excursions, is something I shall always treasure.

Last, but not least, I wish sincerely to thank both Professor F. H. T. Rhodes and Professor D. V. Ager for their constant co-operation and encouragement. My good friend and colleague for very many years, Mr Brian Simpson, has also given me stalwart support.

The author is grateful to the Director of the Institute of Geological Sciences for permission to adapt figures 12, 21, 22 and 27 from Geological Survey photographs, and to Aerofilms Ltd for permission to adapt figure 54 for one of their photographs.

Figures 13 and 23 are based on Owen et al, 1971. Figure 17 is after Thomas and Thomas, 1956, while figure 20 is after Thomas, 1923. Figures 25 and 26 are after Dixon, *Pembroke Memoir*, 1921. Figure 28 is after George and Kelling in Bassett and Bassett, 1971 and D. G. Jones, 1969, while figure 31 is after Trotter, 1948.

Figures 39 and 40 are after O. T. Jones and W. J. Pugh, 1949 while figures 41 and 43 are based on Garwood and Goodyear, 1919.

Figures 42, 74 and 78 are based on University of Wales theses by C. M. Davies, R. Gomm and D. B. Norris, respectively. Figures 44, 47, 49 and 51 are after George, 1940. Figure 58 is based in part on T. M. Thomas in Bassett and Bassett, 1970, while figure 61 is after Coase, 1967. Figure 64 is based on North, Norris and others, while figure 72a is based on Owen et al *Sheffield Congress Excursions*, 1971. Figure 79 is after Squirrell in Bassett and Bassett, 1971 and figure 81 after Squirrell and Downing, *Newport Memoir*.

Figures 65, 66 and 68 are based on Owen et al, 1965 and on Owen and Rhodes, *Geologists' Association Guide No 17*, 1969. Figures 86 and 87 are after T. M. Thomas, 1968 and figure 88 after the same author, 1961.

Index